LEARN ROBOTICS with microcontroller

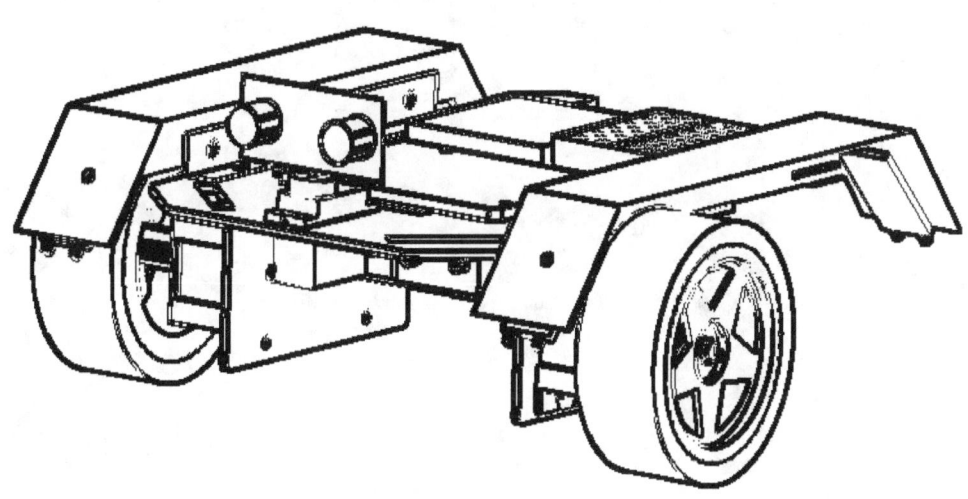

Disclaimer

Table of Contents

Introduction to Robotics

Robotics is not a single subject—it is a meeting point where many branches of engineering come together to solve real problems. When you build even a simple robot, you are working with mechanical structures, electrical power, electronic circuits, sensors, motors, and software logic at the same time. As robots become more capable, decision-making logic and basic forms of machine intelligence are added on top of these foundations.

What makes robotics exciting is that everything you learn has a visible outcome. When your code changes, the robot moves differently. When a sensor is adjusted, the robot reacts in a new way. This direct cause-and-effect relationship makes robotics one of the most practical ways to learn engineering concepts. Many beginners are curious about questions like: How does a robot know where to go? How does it avoid obstacles? Why are robots used instead of humans in some tasks? This learning kit is designed to answer those questions through hands-on construction. Instead of starting with theory alone, you will learn by building, testing, failing, fixing, and improving a working robot.

This kit is suitable for self-learners, engineering students, and technically curious individuals who want to understand robotics from the ground up—not just how to assemble parts, but how and why each part works.

Understanding Different Types of Robots

Robots can be classified based on how much decision-making power they have and how they receive instructions. Some machines perform a single task repeatedly using fixed mechanical or electrical logic. These systems follow the same steps every time and do not adapt to changing conditions. This approach is often called fixed or hard automation. Such machines are extremely reliable for repetitive work but cannot respond intelligently to unexpected situations.

More advanced robots use programmable logic. In these systems, behavior is controlled by software rather than fixed wiring alone. The robot follows a set of rules written in code and can respond differently depending on sensor inputs. For example, if an obstacle is detected, the robot may stop, turn, or choose a new path. This ability to react to conditions makes programmable robots far more flexible. Some robots operate completely on their own once powered on. These are commonly described as autonomous systems. They sense their environment, make decisions locally, and act without human intervention during operation.

Other robots combine internal decision-making with external control. These systems can accept commands from a human operator while also providing feedback about their surroundings. For instance, a remotely controlled vehicle may still use onboard sensors to prevent collisions or maintain stability. This hybrid approach is often referred to as semi-autonomous operation. Understanding these categories helps you design robots appropriately. A robot meant to patrol a table edge needs very different logic from one meant to follow a line or respond to light.

Where Robots Are Used Today

Robots are no longer limited to factories or research labs. They are used in environments that are dangerous, repetitive, or physically demanding for humans. Manufacturing, healthcare assistance, environmental monitoring, agriculture, logistics, and home automation are just a few examples.

At home, simple robotic systems already handle tasks such as cleaning, monitoring, and basic automation. In industry, robots weld, paint, assemble, and inspect with high precision. In exploration, robots operate in places humans cannot safely reach. As technology becomes more accessible, small educational robots play an important role in skill development. They allow learners to experiment with real systems at low cost and low risk.

Microcontroller Robot Project

In this project we are building a differential drive robot that can perform following tasks based on the program.

- Move and rotate forward, backward, right, left in autonomous navigation.

- Determine the surrounding obstacles.

- Avoid obstacles.

- Edge detection and avoidance.

- Can run on a table or similar surface.

- Light detection for light avoidance.

- Table guard mode.

- Line follower mode.

- Communication using light and sound.

Required equipments for Microcontroller Robot Project

Following is the list of equipments that are required for constructing Microcontroller Robot. These equipments are sub-divide into mechanical, electrical and programmable categories.

Mechanical equipments -

1. Robot chassis [3D printed] – 1 set

2. 3mm nut and bolt – 50

3. drive wheel – 2

4. Caster wheel – 1

Electrical & electronics components

5. DC motors - 2

6. Batteries 9 volt - 4

7. Battery clip - 4

8. Servo Motor - 1

9. Ultrasonic sensors - 1

10. Infrared - Ray Sensor - 6

11. Microcontroller Uno R3 - 1

12. Bread board - 1

13. Male to male connection wire

14. Male to female connection wire

15. Motor Controller - 1

16. voltage regulator 7805 - 1

17. LED   - 10

18. Light sensor - 1

19. Sound Buzzer - 1

20. Register 220 ohms - 10

21. Register 4.6 k ohm - 10

22. Capacitor 0.1uf - 2

23. Capacitor 100uf - 2

24. Push switch - 5

25. Relay - 1

Programming tools –

26. Microcontroller IDE

27. Circuit design & simulator.

3D Printed Mechanical Components

The robot kit has the following mechanical parts that need to be 3D printed. All the files mentioned below can be download from the companion site of this book.

☐robot platform

☐Motor Casing

☐Side skirt

☐robot wheel

☐Caster wheel base

☐Edge sensor angle

☐Edge sensor holder

☐ultrasonic sensor holders

☐Battery Holder

Microcontroller Robot Assembly

After printing the above mentioned materials using a 3D printer, the mechanical parts of the robot need to be assembled following the instructions in below

✓ ☐Start with the DC motor.

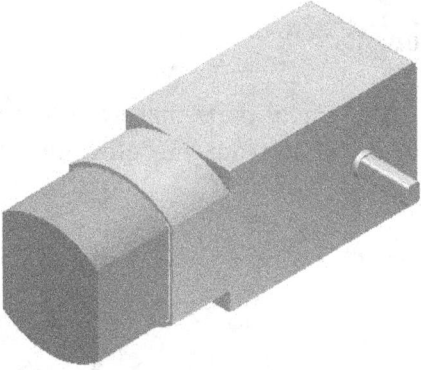

We attach the motor casing and side skirt to the DC motor as shown below. After connecting with the nut bolt, we attach the wheel as shown in following diagram.

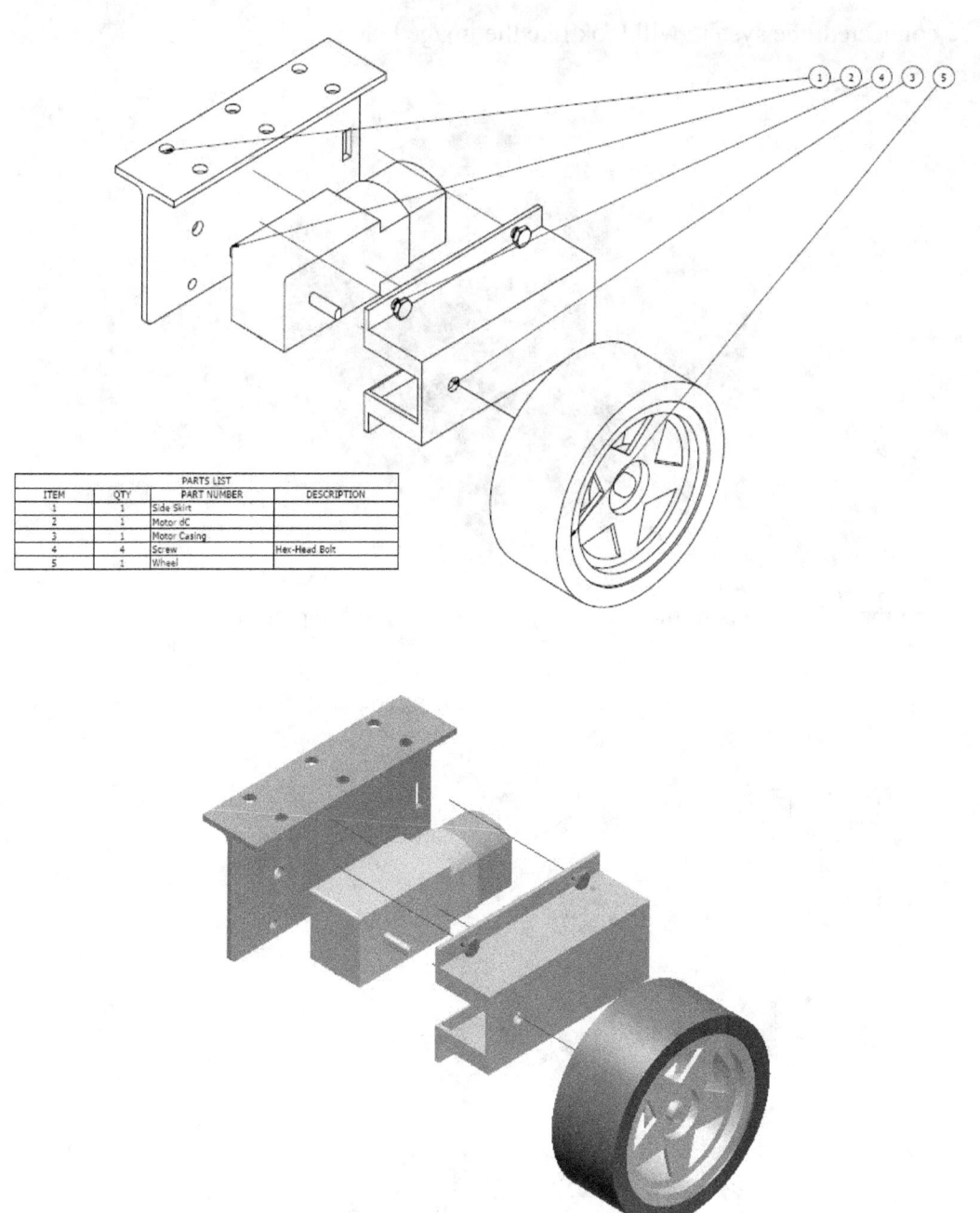

PARTS LIST			
ITEM	QTY	PART NUMBER	DESCRIPTION
1	1	Side Skirt	
2	1	Motor dC	
3	1	Motor Casing	
4	4	Screw	Hex-Head Bolt
5	1	Wheel	

Once connected, the system will look like the image below.

Let's add the side wheels to the chassis base as shown in the picture.

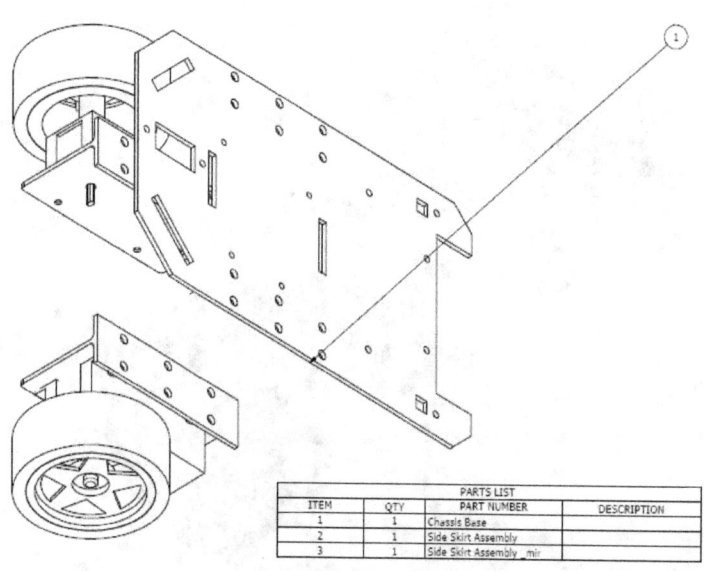

PARTS LIST			
ITEM	QTY	PART NUMBER	DESCRIPTION
1	1	Chassis Base	
2	1	Side Skirt Assembly	
3	1	Side Skirt Assembly _mir	

Once connected, the system will look like the image below.

Now let's place the caster wheel and brush base with the caster bay is as shown below.

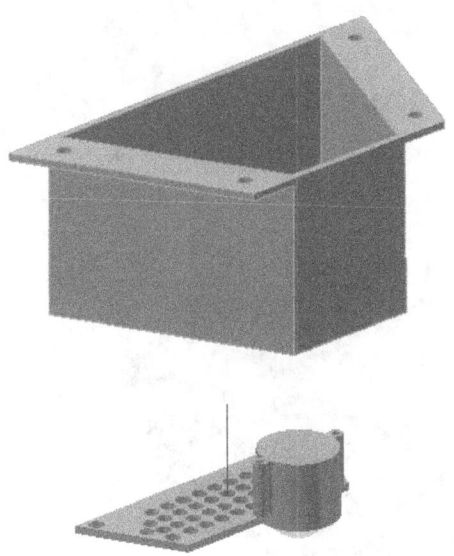

We attach the caster base to the chassis base as shown in the figure.

The connection termination system will look like the image below.

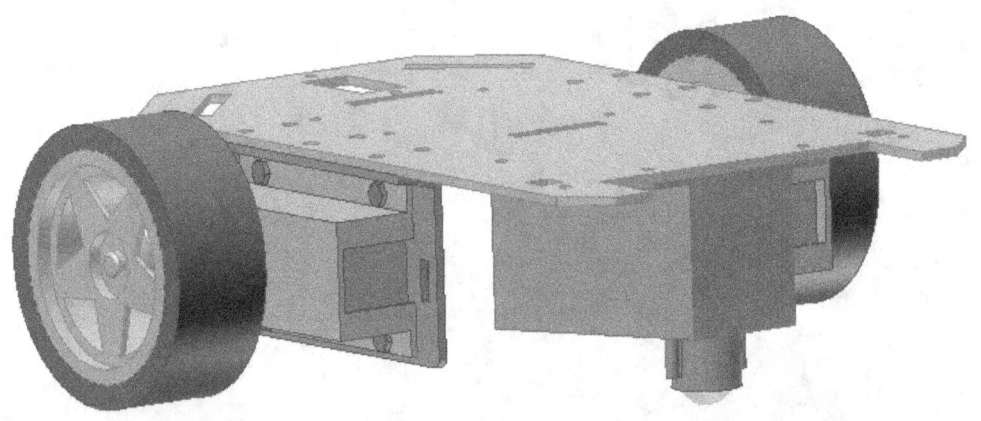

Then we attach the edge sensor angle and the edge sensor holder to the chassis base. Immediately add the battery holder below.

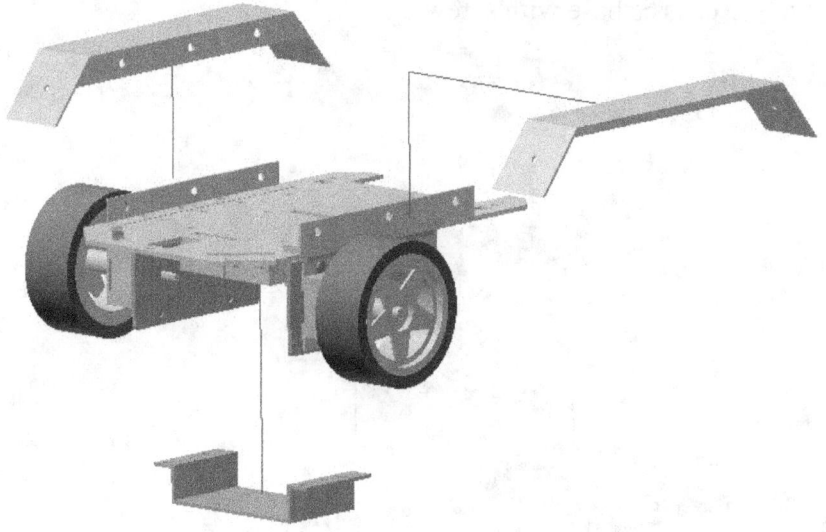

At the end of the work the robot will look like the image below.

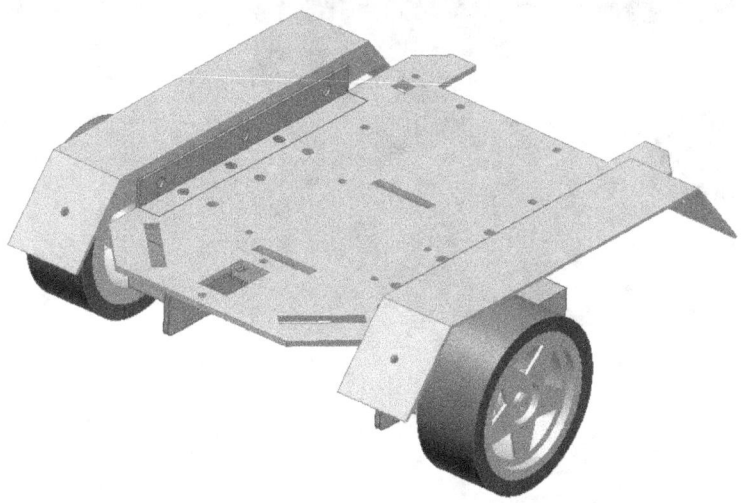

We then add the robot's electronics components, the Microcontroller Uno, the breadboard, the motor controller, and finally the servomotor at the bottom.

Attach the servomotor to the chassis base with screws.

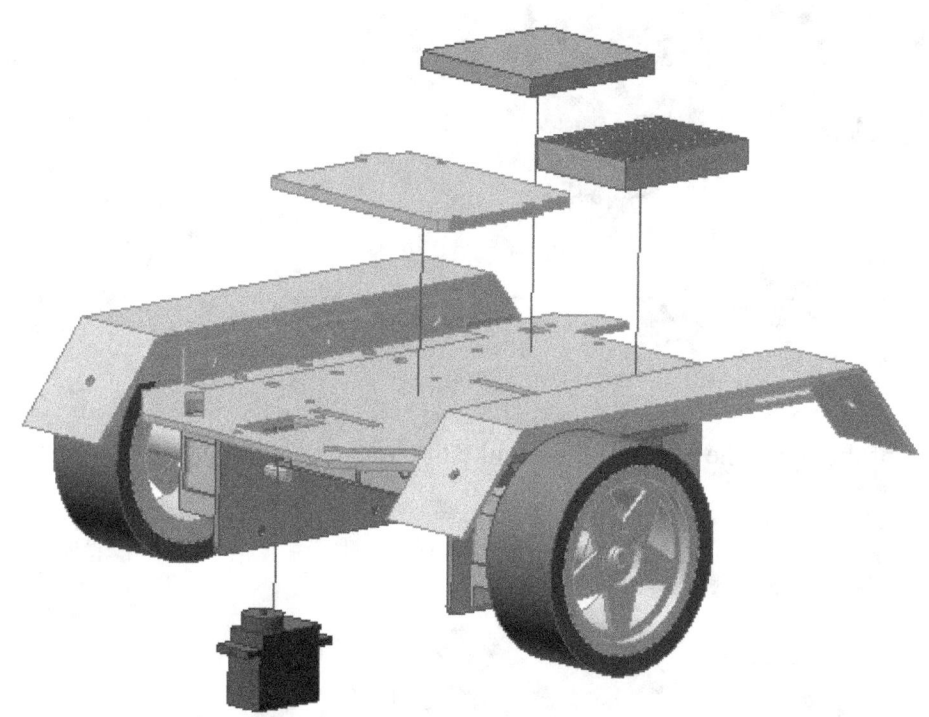

Then install ultrasonic sensor on top of servo motor and infra sensor at the skirt.

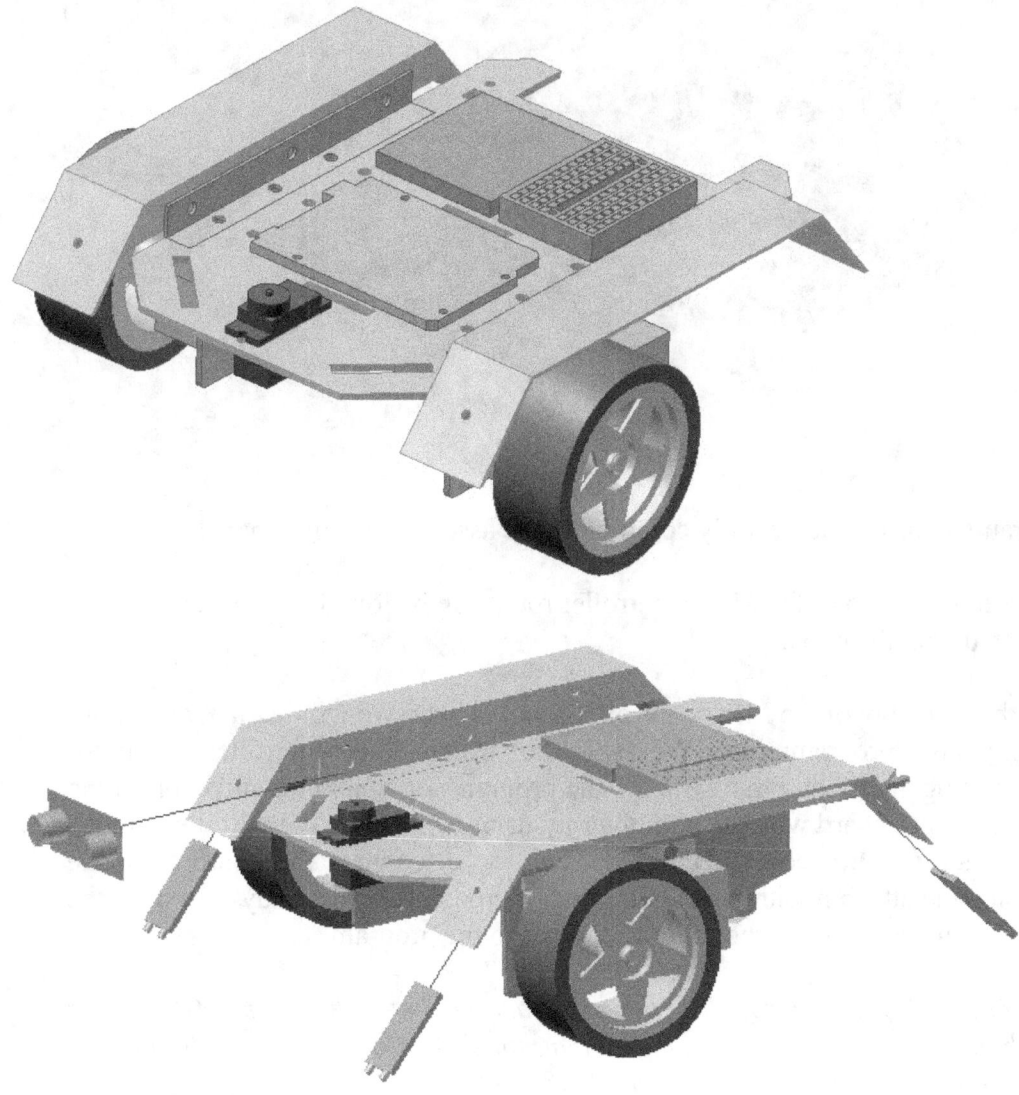

When the assembly is complete, the robot will look like the image below.

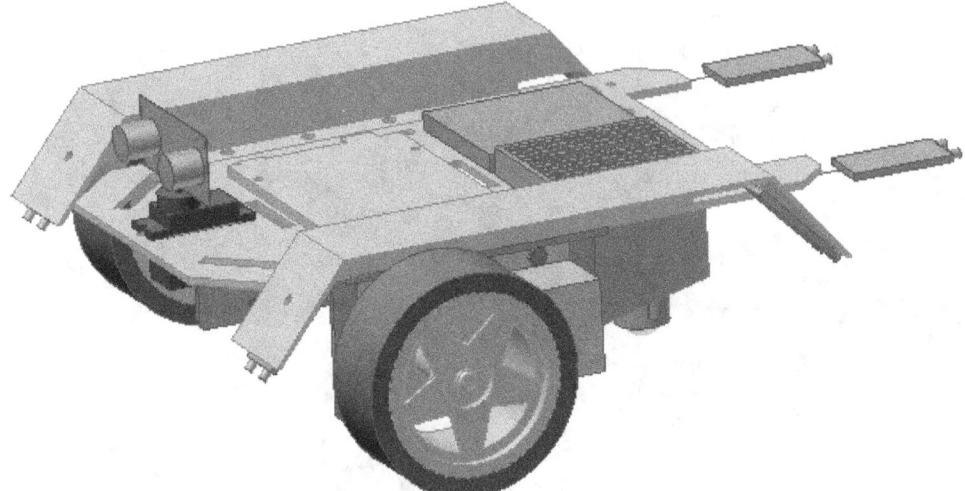

Congratulations on successfully completing the assembly of the robot !!

The assembly video of the Microcontroller robot can be found on the site associated with the book.

After the assembly of the Microcontroller robot is over, we will learn about circuit building and programming. We will be using Microcontroller Uno for programming the robot. First we write the programs on computer and connect the Microcontroller board with the computer by using USB port. We will then upload the program to Microcontroller board. Then we disconnect the board from the computer and attach it with our robot. Now Microcontroller is ready to control the robot's circuit and operate the robot according to the program.

*** Caution - The power switch of the robot and motor controller must be turned off while uploading the program. At that time, Microcontroller will only be powered by power from the computer's USB port.*

Connecting Microcontroller to a computer

Before building a circuit using Microcontroller, we need to know a few things. So first we need to know about the hardware and software requirement to run Microcontroller module. Then we take a simple check on our Microcontroller Learning Board to see if it works. So let's get started.

Hardware

For this robot project, we will use a commonly available microcontroller board based on the "Uno R3" design. This particular board layout is widely used in education because it is simple, reliable, and well supported by learning resources. Many other boards exist with different features, but starting with this one keeps the learning curve manageable and helps beginners focus on understanding core concepts rather than hardware complexity.

This board is built around an open hardware design. That means its circuit layout and behavior are publicly documented, and compatible versions are available from many manufacturers. From a learner's perspective, this is useful because it removes dependency on a single supplier and encourages experimentation.

If you look closely at the board, you will notice clearly labeled sections: connection pins, power inputs, indicator lights, and the main processing chip. Each of these parts plays a specific role. The board acts as the central decision-maker of the robot. It does not move motors or sense the environment by itself—instead, it coordinates everything by sending and receiving electrical signals.

Inputs and Outputs

To understand how a robot works, it helps to think in terms of information flow. The controller board constantly receives information from the outside world and sends commands back out.

Information coming *into* the board is called input. Inputs usually come from sensors, such as distance sensors, light sensors, or switches. These sensors convert physical conditions into electrical signals that the board can read.

Information sent *out* from the board is called output. Outputs are used to control devices like motors, indicator lights, buzzers, or relays. When the board decides to move forward, turn, or stop, it does so by changing the output signals sent to motor drivers or other control circuits.

The board used in this project provides fourteen digital connection pins. Each of these pins can be configured in software to act either as an input or an output. This flexibility is extremely important. The same physical pin can read a sensor in one project and control a motor in another, depending entirely on how the program is written.

A common beginner mistake is to assume that pins are permanently fixed as inputs or outputs. In reality, the role of each pin is defined by the program, not by the hardware alone.

Powering the Controller Board Safely

The controller board needs electrical power to operate, and there are several safe ways to supply it. When the board is connected to a computer using a USB cable, it receives power directly from the computer. This method is very useful during programming and testing. The same cable is also used to transfer the control program from the computer to the board's memory. Once the program is uploaded, the board will remember it even after the cable is disconnected.

For standalone operation—when the robot is moving independently—a separate power source is required. This is typically done using a battery or an external power adapter rated around nine volts. When using a battery, it is important to ensure correct polarity and a secure connection. Loose or reversed connections are a frequent cause of non-working robots.

When using an external adapter connected to a wall outlet, always double-check the voltage rating. Supplying too much voltage can permanently damage the board. Supplying too little voltage may cause unpredictable behavior, such as random resets or weak motor performance.

A good practice during early testing is to power the board through the USB connection first. Once the program works correctly, switch to battery or adapter power for full robot operation.

Using a Breadboard for Circuit Building

The breadboard is one of the most important tools for learning electronics. It allows you to build and modify circuits without soldering, which means mistakes can be corrected easily and safely.

Physically, a breadboard is a plastic block filled with evenly spaced holes. Under the surface, these holes are connected in specific patterns using metal strips. Although this is not visible from the outside, understanding these internal connections is critical.

Rows of holes are connected together horizontally or vertically depending on their position on the board. Power rails run along the edges and are typically used to distribute voltage and ground throughout the circuit. The central area is used for connecting components such as resistors, sensors, and wires. A common beginner error is assuming that all holes are connected together. They are not. Always take time to understand which holes share an electrical connection before inserting components.

Because no soldering is required, breadboards are ideal for testing ideas, debugging circuits, and learning how components interact. Once a circuit works reliably on a breadboard, it can later be converted into a permanent version if needed.

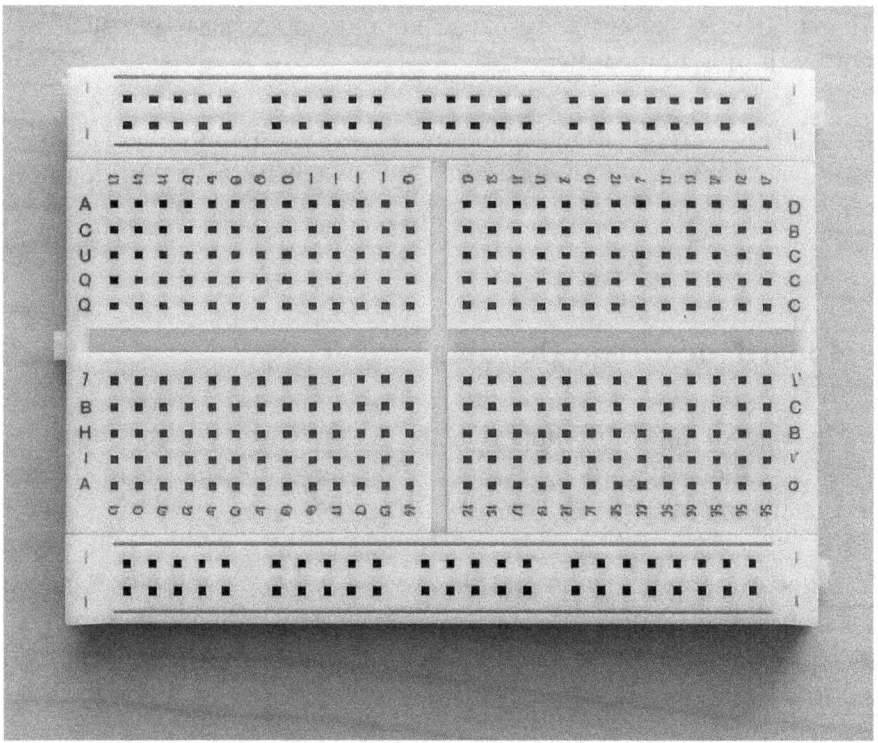

On above diagram, the red and blue lines are the power line of the board. The holes in this line are connected in parallel. Positive power is supplied on the red line and negative power is supplied on the blue line.

Jumper Wires and Reliable Connections

Jumper wires are simple but critical components in any breadboard-based project. Their job is to create temporary electrical connections between different points in a circuit without soldering.

Each jumper wire has a metal pin at the end. When this pin is inserted into a breadboard hole, it is held tightly by an internal spring contact. This pressure ensures a stable electrical connection while still allowing the wire to be removed or repositioned easily.

Good connections matter more than most beginners realize. A loose jumper wire can cause intermittent failures that look like software bugs but are actually hardware problems. If a robot behaves inconsistently, always check jumper wires first. Slightly wiggling a wire while observing behavior is a simple way to detect poor contact.

It is also good practice to keep wiring short, neat, and well organized. Long or tangled wires increase the chance of accidental disconnections and make troubleshooting harder.

Programming the Controller Board

To make the robot behave intelligently, we must provide instructions to the controller board in the form of a program. These instructions define how the robot responds to sensor input and how it controls motors, lights, and sound.

Programs are written using a software tool known as an Integrated Development Environment, commonly abbreviated as IDE. In this project, we use the official development software associated with the Microcontroller ecosystem. This software runs on widely used operating systems and is designed to be approachable for beginners.

A program written for the board is often referred to as a *sketch*. The term simply means a small, focused program that performs a specific task. Even though the name sounds informal, sketches can control surprisingly complex systems.

After writing a sketch, it is transferred from the computer to the controller board using a USB cable. The same cable also supplies power during programming and testing. Once uploaded, the board stores the program in memory and runs it automatically whenever power is applied.

Writing, Checking, and Uploading Code

The workflow for programming follows a predictable sequence.

First, the code is typed or pasted into the editor window of the IDE. All example programs used in this book are provided through the companion resource site, so you do not need to write everything from scratch.

Before sending the program to the board, it is important to check it for errors. The IDE includes a verification step that scans the code for mistakes such as missing symbols, incorrect structure, or undefined variables. This step prevents many common problems before they reach the hardware.

If the code passes verification, it can then be uploaded to the board. During upload, indicator lights on the board may blink briefly. Once the process completes, the board immediately begins running the new program. If the upload fails, common causes include a loose USB cable, an incorrect board selection in the software, or another program already using the communication port.

Understanding Libraries and Why They Matter

As projects grow more complex, writing everything from scratch becomes inefficient. This is where libraries come in. A library is a collection of pre-written code that performs a specific function, such as reading a sensor, controlling a display, or handling input from a keypad. Libraries allow you to reuse proven logic instead of reinventing it each time.

For this robot project, certain libraries are required. These libraries are provided through the companion site and are selected to simplify learning rather than hide important details. To install a library using the IDE, you select the option to add a compressed library file, then point the software to the downloaded file. After installation, restarting the IDE ensures the new library is recognized.

Libraries can also be installed manually by extracting the files and placing them in the designated library folder on your system. This method is useful if automatic installation fails or if you want more control over file organization.

At the top of a sketch, libraries are referenced using a special include statement. This tells the compiler to make the library's functions available to your program. If a required library is missing, the code will not compile.

Installing the Development Software

The development software used in this project is available freely from the official source associated with the platform. During installation, selecting default options is recommended for beginners, as this ensures all required drivers and tools are included. After installation completes, connecting the controller board to the computer with a USB cable will power the board and establish communication. A small indicator light on the board will turn on, confirming that power is present.

At this point, the system is ready for programming and testing.

Then open the IDE software on the computer and write or load the program.

```
// ===== Simple LED Blink Demo =====
// This sketch controls an LED by turning it
// ON and OFF continuously using a digital pin.

// Define the output pin where the LED is connected
const int indicatorPin = 10;

void setup() {
  // Configure the selected pin as an output pin
  pinMode(indicatorPin, OUTPUT);

  // Ensure the LED starts in OFF state
  digitalWrite(indicatorPin, LOW);
}

void loop() {
  // Send HIGH signal to power the LED
  digitalWrite(indicatorPin, HIGH);

  // Wait while the LED stays ON (0.9 seconds)
  delay(900);
```

```
// Cut power to the LED
digitalWrite(indicatorPin, LOW);

// Pause while the LED remains OFF (0.5 seconds)
delay(500);
}
```

Then select

Tools> Board and select the board from the menu

Let's select the serial port

Here is a clean, original rewrite that improves clarity, avoids copied phrasing, and explains the idea in a practical, beginner-friendly way. The tone is hands-on and instructional, based on real usage rather than textbook wording.

When uploading a program to the controller board, the computer communicates with it through a communication port. On many systems, ports with lower numbers such as COM1 or COM2 are already assigned to built-in hardware functions. For this reason, the board is often assigned a higher-numbered port, commonly something like COM3 or above. However, relying on a specific number is not a reliable method.

A simple and foolproof way to identify the correct port is to observe the list of available ports before and after connecting the board. Disconnect the board and note which port disappears from the list. Then reconnect it and see which port appears again. The newly appearing port is the one currently linked to the board.

Once the correct port is selected in the development software, the program can be transferred to the board. To begin the upload, click the upload button located near the top-left area of the interface. After clicking it, wait patiently for a few seconds. During the upload process, small indicator lights labeled for data transmission and reception will blink. These lights show that data is actively moving between the computer and the board. When the process completes successfully, a confirmation message will appear indicating that the upload is finished.

If this message does not appear, common issues include selecting the wrong port, using a faulty cable, or having another application already using the same communication channel. Checking these points usually resolves the problem quickly. Once uploaded, the board immediately begins running the new program, even if the cable is later disconnected and power is supplied from another source.

If you want to add a library file, click Sketch> Include Library> Add ZIP Library

Let's browse and show the file location

Basic Microcontroller programming

After successfully connecting the board to the computer, we can begin learning the basics of programming. Every program written for this platform follows a simple and consistent structure made up of two main parts. The first part is the `void setup()` function. This section runs only once, immediately after the board is powered on or reset. It is used to define initial settings and perform tasks that do not need to be repeated.

The second part is the `void loop()` function. After the setup phase is complete, the code inside this function runs continuously. It keeps repeating for as long as the board remains powered, allowing the program to perform ongoing actions. To understand this structure more clearly, we will look at a simple example.

Project Construction

We are already familiar with the hardware and software involved. Now we will build a very basic project. In this project, an LED will be turned on and off using a program. By completing this simple task, we will develop a clearer understanding of how the hardware and software work together.

We have already learned that a sketch is simply a program that gives instructions to the board. One important point to remember is that the board can store only one sketch in its memory at a time. After a sketch is uploaded from the computer, that same program will continue to run every time the board is powered on, until a new sketch replaces it.

For this project, we will use the default example program available in the IDE to control an LED. This program switches the LED on for one second and then turns it off for the next second. While working with LEDs, it is important to understand that they only light up when current flows in the correct direction. If the polarity is reversed, the LED will not turn on. The longer leg of the LED, or the leg with the larger internal plate, must be connected to the positive side of the circuit.

LEDs are designed to work with a small amount of current. If too much current passes through them, they can be damaged permanently. To prevent this, a resistor is always used with an LED. In this case, the board already includes a built-in resistor connected to digital pin 13, which makes it safe and convenient for this basic experiment.

Now follow the steps below to build the project.

First, insert the longer leg of the LED into pin 13 on the board and connect the shorter leg to the GND pin.
Second, connect the board to the computer using a USB cable.
Third, copy the given sketch into the IDE.
Fourth, click the verify button so the IDE can check whether the program has any errors.
Fifth, click the upload button to transfer the program to the board.

Sketch Review

In this section, we will explain each part of the sketch.

First, lines that begin with // are comments. These lines are meant only for the reader and are ignored by the board. Comments are useful for writing notes or explanations inside the program. If multiple lines of comments are needed, they can be written between /* and */. Anything inside these symbols is ignored during execution.

Second, pin number 13 is assigned a name, such as LED. This means that whenever this name is used in the program, it refers to pin 13.

Third, curly brackets are used to group instructions together. Any code written between a pair of brackets belongs to the same section.

Fourth, pin 13 is configured as an output because it will supply power to the LED. After this instruction, the setup section ends.

Fifth, the loop section begins here. Once the loop starts running, the instructions inside it execute once and then repeat continuously as long as the board has power.

Sixth, the program sets the state of pin 13 to HIGH. This allows current to flow from the pin to the LED, turning it on.

Seventh, the program pauses for one second. Time delays are measured in milliseconds, so one second equals one thousand milliseconds.

Eighth, the state of pin 13 is set to LOW. This stops the current flow and turns the LED off.

Ninth, the program waits for another one-second delay.

Tenth, the loop section ends here. After this point, the program automatically returns to the beginning of the loop and repeats the same steps. Curly brackets are very important in a program. Missing or misplaced brackets often cause errors during compilation.

When this program runs, the LED will continuously turn on and off.

At this stage, you have learned how to write and upload a basic program. Next, we will begin exploring the components used in the project. Detailed descriptions and images of these components are provided in the appendix section at the end of the book.

Writing Programs for Robots

At this stage, the robot has already been assembled. Now that we understand the basics of the controller board and programming, we can begin writing programs that make the robot move. We will start with a simple and practical task. The robot is equipped with two DC motors, and in this project we will control those motors through code so the robot can perform basic movements.

Before building the physical circuit, we will first create a simulation for each project. A simulation circuit is a digital version of the real circuit. It is drawn using computer software, and the same components found in the actual hardware are connected virtually. These simulated circuits can be run and tested on a computer, allowing us to observe behavior and detect mistakes early. This approach helps identify wiring errors, logic problems, or incorrect connections before assembling the real circuit. Here, we will create the first simulation for controlling DC motors.

To build these simulations, we will use an online circuit design and simulation tool provided by Tinkercad® software. A step-by-step guide for using this tool is available through the companion resources for this book. Since the tool runs in a web browser, it can be accessed by visiting its official website and signing in.

Interfacing the Drive Motors with the Robot

The robot uses two DC motors as its primary drive system. These motors are responsible for all movement, including forward motion, turning, and rotation. In this project, we will connect the motors to the controller board and control them entirely through software.

With two independently controlled motors, the robot can perform several different movements. When both motors rotate forward at the same time, the robot moves straight ahead. When both rotate in reverse, the robot moves backward. If one motor rotates forward while the other remains stopped, the robot turns in the direction of the stationary motor. When one motor rotates forward and the other rotates backward, the robot spins in place. If both motors are stopped, the robot remains stationary and effectively brakes.

All of these movements are achieved by controlling the electrical signals sent to the motors. These signals are generated by the program running on the controller board.

Before writing the code, it is important to understand the motor interface circuit. The robot uses two six-volt DC motors, which cannot be driven directly from the controller pins. Instead, a motor driver integrated circuit is used between the controller and the motors. This driver receives low-power control signals and safely switches higher current to the motors. By sending different signal combinations to the driver, we can control motor direction and motion.

The pin layout of the motor driver determines how signals are connected and how each motor responds. Understanding this connection is essential before moving on to programming, as the code directly corresponds to how these pins are wired.

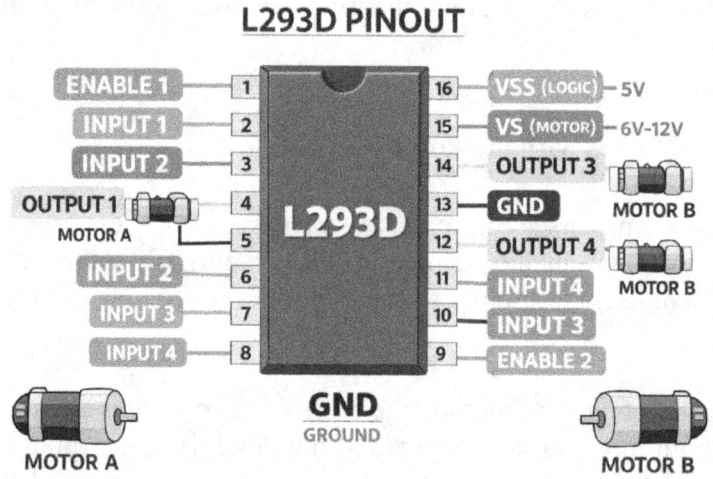

If we examine the pin layout of the IC, we can see that it has a total of sixteen pins. Among these, pins 4, 5, 12, and 13 are connected to ground. These ground pins not only complete the electrical circuit but also help release heat generated inside the IC during operation, which improves stability and protects the device.

Pins 1 and 9 function as enable pins. These pins control whether the motors are allowed to run. When an enable pin is supplied with the appropriate signal, the corresponding motor channel becomes active. If the enable pin is disabled, the motor connected to that section will stop, regardless of the input signals.

The input pins are used to send control signals to the IC, while the output pins are connected directly to the motors. By changing the signal pattern applied to the input pins, the direction in which a motor rotates can be controlled. Reversing the input signals reverses the motor's rotation.

A single IC can control two separate DC motors. The first motor is operated using output pins 1 and 2, while the second motor is operated using output pins 3 and 4. This arrangement makes it possible to control both motors independently using one motor driver IC.

Note that a voltage controller IC7805 has been used in the above circuit. This IC can provides 5 volt output, which we can supply to different parts of the circuit.

Below is a motor control program for robots. Some comments are made near the line to make the code more understandable.

```
/*
  Dual DC Motor Test Sketch

  This program demonstrates basic motion control
  using two DC motors and a motor driver.

  NOTE:
  If a motor spins in the wrong direction, you can fix it by:
  1) Reversing the motor wires, OR
  2) Swapping HIGH and LOW logic in the direction functions.
*/

//                        ========================          Pin          Assignment
========================
// Direction control pins for left motor
const int LM_DIR1 = 8;
const int LM_DIR2 = 9;

// Direction control pins for right motor
const int RM_DIR1 = 12;
```

```cpp
const int RM_DIR2 = 13;

// Enable pins (can be PWM-capable for speed control)
const int LM_ENABLE = 10;
const int RM_ENABLE = 11;

// ======================= Initialization =======================
void setup() {
  // Configure motor direction pins as outputs
  pinMode(LM_DIR1, OUTPUT);
  pinMode(LM_DIR2, OUTPUT);
  pinMode(RM_DIR1, OUTPUT);
  pinMode(RM_DIR2, OUTPUT);

  // Configure enable pins
  pinMode(LM_ENABLE, OUTPUT);
  pinMode(RM_ENABLE, OUTPUT);

  // Power up both motor channels
  digitalWrite(LM_ENABLE, HIGH);
  digitalWrite(RM_ENABLE, HIGH);

  // Ensure motors are not moving at startup
  haltAllMotors();
}

// ======================= Low-Level Motor Control =======================
// Left motor control states
void leftMotorForward() {
  digitalWrite(LM_DIR1, HIGH);
  digitalWrite(LM_DIR2, LOW);
}

void leftMotorBackward() {
  digitalWrite(LM_DIR1, LOW);
```

```cpp
  digitalWrite(LM_DIR2, HIGH);
}

void leftMotorIdle() {
  digitalWrite(LM_DIR1, LOW);
  digitalWrite(LM_DIR2, LOW);
}

// Right motor control states
void rightMotorForward() {
  digitalWrite(RM_DIR1, HIGH);
  digitalWrite(RM_DIR2, LOW);
}

void rightMotorBackward() {
  digitalWrite(RM_DIR1, LOW);
  digitalWrite(RM_DIR2, HIGH);
}

void rightMotorIdle() {
  digitalWrite(RM_DIR1, LOW);
  digitalWrite(RM_DIR2, LOW);
}

//          ========================          Motion          Commands
========================
void driveStraightForward() {
  leftMotorForward();
  rightMotorForward();
}

void driveStraightBackward() {
  leftMotorBackward();
  rightMotorBackward();
}
```

```
// Soft turns (one motor active, one stopped)
void curveLeftForward() {
  leftMotorForward();
  rightMotorIdle();
}

void curveRightForward() {
  leftMotorIdle();
  rightMotorForward();
}

void curveLeftBackward() {
  leftMotorBackward();
  rightMotorIdle();
}

void curveRightBackward() {
  leftMotorIdle();
  rightMotorBackward();
}

// Rotation on the spot
void rotateLeft() {
  leftMotorForward();
  rightMotorBackward();
}

void rotateRight() {
  leftMotorBackward();
  rightMotorForward();
}

// Stop all motion (coast mode)
void haltAllMotors() {
  leftMotorIdle();
  rightMotorIdle();
```

```
}
```

```
void loop() {
  // Move forward
  driveStraightForward();
  delay(2000);

  haltAllMotors();
  delay(300);

  // Move backward
  driveStraightBackward();
  delay(2000);

  haltAllMotors();
  delay(300);

  // Rotate clockwise
  rotateRight();
  delay(2000);

  haltAllMotors();
  delay(300);

  // Rotate counter-clockwise
  rotateLeft();
  delay(2000);

  // Longer pause before repeating sequence
  haltAllMotors();
  delay(800);
}
```

If the above simulation circuit and the code work properly, we will be able to build a real circuit of the robot, same as simulation circuit. However, many of the components in the simulation may be slightly different from the actual circuit.

Circuit Construction

Assemble the circuit components as shown in the image above and upload the program to Microcontroller. Please note that, the robot's battery power must be turned off when uploading the program to Microcontroller. The Microcontroller will only be powered by a USB drive while uploading the program.

Ultrasonic sensor Simulation for obstacle detection

Previously, we connected the DC motors to the robot. In this section, we will add an ultrasonic sensor so the robot can detect obstacles in front of it. The program written here is intended for simulation, so you may notice small differences between the simulation code and the code used in the real circuit.

The main reason is the sensor pin configuration. In the simulation, a three-pin ultrasonic sensor is used, where the same signal pin is used for both transmitting and receiving. In a real build, a four-pin ultrasonic sensor is commonly used, where one pin is dedicated to sending the ultrasonic pulse and another pin is dedicated to receiving the return signal. Because of this, the simulation requires a three-pin program, while the physical circuit requires a four-pin program. Here, we will write the program for the three-pin version.

An ultrasonic sensor works using two parts. One part produces a sound pulse at an ultrasonic frequency, and the other part listens for the reflected sound. The program measures the time gap between sending the pulse and receiving the echo. That time value is then converted into distance, allowing the system to estimate how far an object is from the sensor.

Since this method depends on reflected sound, there are situations where the sensor may not behave as expected. If the sound hits a surface that absorbs the pulse instead of reflecting it, the echo may not return. Similarly, if the pulse strikes an angled surface or a corner, it may bounce away from the sensor rather than back toward it. Even with these limitations, the sensor works well for detecting most obstacles placed directly in front of it.

In this program, the blue indicator light will remain on and the red indicator light will remain off during normal conditions. If the sensor detects an obstacle within the set range, the program will switch the indicators: the blue light will turn off and the red light will turn on.

This program can be used in simulations.

```
// ===== Ultrasonic Obstacle Indicator =====
// Uses one ultrasonic signal pin and two LEDs.
// - Red LED lights up when an object is nearby
// - Blue LED lights up when the path is free

const int ultraIO  = 7;  // shared trigger/echo pin
const int ledAlert = 2;  // red warning LED
const int ledClear = 3;  // blue safe-status LED

void setup() {
  // Initialize serial output for distance readings
  Serial.begin(9600);

  // Configure LED pins as outputs
  pinMode(ledAlert, OUTPUT);
  pinMode(ledClear, OUTPUT);
}

void loop() {
  long pulseDuration;  // stores echo pulse time
  long rangeCm;        // calculated distance in centimeters

  // Prepare ultrasonic pin to transmit trigger pulse
  pinMode(ultraIO, OUTPUT);
  digitalWrite(ultraIO, LOW);
  delayMicroseconds(2);

  // Send a short ultrasonic trigger
  digitalWrite(ultraIO, HIGH);
  delayMicroseconds(5);
  digitalWrite(ultraIO, LOW);

  // Switch pin to input mode to capture echo signal
  pinMode(ultraIO, INPUT);
```

```
pulseDuration = pulseIn(ultraIO, HIGH);

// Convert time duration to distance
rangeCm = microsecondsToCm(pulseDuration);

// Print distance value to Serial Monitor
Serial.print("Measured distance: ");
Serial.print(rangeCm);
Serial.println(" cm");

// LED indication logic
if (rangeCm < 100) {
  // Object detected nearby
  digitalWrite(ledAlert, HIGH);
  digitalWrite(ledClear, LOW);
} else {
  // Path is clear
  digitalWrite(ledAlert, LOW);
  digitalWrite(ledClear, HIGH);
}

// Small delay to stabilize readings
delay(100);
}

// Converts echo time (microseconds) into distance (cm)
long microsecondsToCm(long echoMicroseconds) {
  return echoMicroseconds / 29 / 2;
}
```

Interface ultrasonic sensor for obstacles detection

The ultrasonic sensor used for obstacle detection has four separate pins. One of these pins is configured as the trigger pin, which is responsible for sending out the ultrasonic pulse. Another pin is configured as the echo pin, which receives the reflected signal after it bounces back from an object.

In this program, the blue indicator light remains on and the red indicator light stays off under normal conditions. When the sensor detects an obstacle within the defined range, the program responds by turning the blue light off and switching the red light on.

This program can be used by uploading to Microcontroller.

```
// ===== Ultrasonic Distance Alert System =====
// Uses an ultrasonic sensor with separate TRIG and ECHO pins.
// - Red LED turns ON when an obstacle is close
// - Blue LED remains ON when the area is clear

const int trigOutPin  = 7;   // pin used to send ultrasonic pulse
const int echoInPin   = 6;   // pin used to receive echo signal
const int ledWarning  = 2;   // red LED for obstacle alert
const int ledSafe     = 3;   // blue LED for clear path indication

void setup() {
  // Start serial communication for debugging and monitoring
  Serial.begin(9600);

  // Configure LED pins
  pinMode(ledWarning, OUTPUT);
  pinMode(ledSafe, OUTPUT);

  // Configure ultrasonic sensor pins
  pinMode(trigOutPin, OUTPUT);
```

```
  pinMode(echoInPin, INPUT);
}

void loop() {
  long echoPulseTime;   // duration of echo pulse in microseconds
  long objectDistance; // calculated distance in centimeters

  // Reset trigger pin before sending pulse
  digitalWrite(trigOutPin, LOW);
  delayMicroseconds(2);

  // Generate ultrasonic trigger signal
  digitalWrite(trigOutPin, HIGH);
  delayMicroseconds(5);
  digitalWrite(trigOutPin, LOW);

  // Read echo pulse length
  echoPulseTime = pulseIn(echoInPin, HIGH);

  // Convert echo time into distance
  objectDistance = echoTimeToCm(echoPulseTime);

  // Output distance value to Serial Monitor
  Serial.print("Measured distance = ");
  Serial.print(objectDistance);
  Serial.println(" cm");

  // Decision logic for LED indicators
  if (objectDistance < 100) {
    // Obstacle detected within warning range
    digitalWrite(ledWarning, HIGH);
    digitalWrite(ledSafe, LOW);
  } else {
    // No obstacle detected nearby
    digitalWrite(ledWarning, LOW);
    digitalWrite(ledSafe, HIGH);
```

```
  }

  // Small delay for stable sensor readings
  delay(100);
}

// Converts ultrasonic echo duration to distance in centimeters
long echoTimeToCm(long echoTimeMicroseconds) {
  return echoTimeMicroseconds / 29 / 2;
}
```

Interface servomotor with the robot for rotating the ultrasonic sensor

Here we will write a program to control a servo motor using the Microcontroller board. We have already worked with the ultrasonic sensor. That sensor can only measure what is directly in front of it and can estimate distance. In the earlier program, we used those distance readings to switch a red and a blue LED on and off. Now, if we want the ultrasonic sensor to check obstacles not only in front, but also on the left and right, we can mount the sensor on top of a servo motor and rotate it from side to side. By turning the servo, the sensor can "look" in different directions.

In this program, we will focus on controlling the servo motor. A typical servo has three wires. One wire connects to positive power, another connects to ground, and the middle wire carries the control signal between the servo and the Microcontroller board. The circuit diagram at the end of the program shows how these connections are made. At the start of the program, we use a header file named `Servo.h`. This file contains ready-to-use functions for controlling servos. The file can be obtained from the companion resources provided with this book.

Next, we create a servo object and define a few variables. A servo motor contains a small DC motor inside, but it is different from an ordinary DC motor because it can hold and report its position. Inside the servo, a feedback device is connected to the output shaft through gears. In the small servo used here, the feedback device is a potentiometer. As the shaft rotates, the potentiometer changes its resistance. The control circuit inside the servo reads that changing resistance and uses it to determine the shaft position. In the program, the variable named `pos` is used to store the target position of the servo shaft, and we set its initial value to 30.

We then write a function called `servoSweep()`, which rotates the servo through a series of positions. This function uses `for` loops to move the servo gradually rather than jumping suddenly. A `for` loop has three parts: the starting value, the condition that decides when the loop stops, and the amount the value changes each time the loop repeats.

In the first sweep section, the loop starts from 30 and increases step by step until it reaches 90. Each time the loop value increases by one, the program sends that new position value to the servo. So the servo moves to 30, then 31, then 32, and continues this way until it reaches 90. This creates a smooth motion from the center area toward one side.

In the next section, the program continues the movement from 90 up to 150. This shifts the servo farther in the same direction, allowing the sensor to scan to the other side. After reaching the far end, the program then moves the servo back from the edge toward the middle again. In the final part, the servo is moved from the middle toward the left side, completing the full scanning motion. Inside the `setup()` function, we only specify that the servo's control wire is connected to pin number 6. In the `loop()` section, we call the `servoSweep()` function. Because `loop()` repeats continuously while the board is powered, the sweep function will run again and again, making the servo move left and right repeatedly.

```
#include <Servo.h>   // library for controlling servo motors

Servo scanServo;     // servo object used for scanning motion

int angle = 30;      // stores current servo angle
int distanceIn = 0;  // placeholder variable (unused)
int distanceCm = 0;  // placeholder variable (unused)

// Performs a left-center-right-center sweep motion
void performScanMotion() {

  // Rotate from left position toward center
  for (angle = 30; angle <= 90; angle++) {
    scanServo.write(angle);   // update servo angle
    delay(15);                // allow time for movement
```

```
}

// Continue rotation from center to right side
for (angle = 90; angle <= 150; angle++) {
  scanServo.write(angle);
  delay(15);
}

// Return from right side back to center
for (angle = 150; angle >= 90; angle--) {
  scanServo.write(angle);
  delay(15);
}

// Move from center back to left side
for (angle = 90; angle >= 30; angle--) {
  scanServo.write(angle);
  delay(15);
  }
}

void setup() {
  // Attach servo signal wire to digital pin 6
  scanServo.attach(6);
}

void loop() {
  // Continuously repeat the scanning motion
  performScanMotion();
}
```

Simulation of connecting servomotor and ultrasonic sensor with robot - 1

The servomotor and ultrasonic sensors are connected together in this program. This program can only be used for simulation work.

```cpp
#include <Servo.h>   // library required for servo control

// ------------------- Hardware pin definitions -------------------
const int ultraPin   = 7;  // shared trigger/echo pin for ultrasonic sensor
const int ledDanger  = 2;  // red LED: obstacle detected
const int ledSafe    = 3;  // blue LED: path is clear

Servo sweepServo;          // servo used for scanning motion

int servoAngle = 30;       // current servo angle
int inchValue  = 0;        // reserved variable (unused)
int cmValue    = 0;        // reserved variable (unused)

// ------------------- Initial setup -------------------
void setup() {
  // Attach servo signal wire
  sweepServo.attach(6);

  // Start serial monitor for distance feedback
  Serial.begin(9600);

  // Configure LED pins
  pinMode(ledDanger, OUTPUT);
  pinMode(ledSafe, OUTPUT);
}
```

```
// -------------------- Servo scanning routine --------------------
// Servo sweeps left → center → right → center,
// taking distance readings at key positions.
void performSweep() {

  // Sweep toward center
  for (servoAngle = 30; servoAngle <= 90; servoAngle++) {
    sweepServo.write(servoAngle);
    delay(15);
  }
  checkObstacle();

  // Sweep toward right side
  for (servoAngle = 90; servoAngle <= 150; servoAngle++) {
    sweepServo.write(servoAngle);
    delay(15);
  }
  checkObstacle();

  // Return back toward center
  for (servoAngle = 150; servoAngle >= 90; servoAngle--) {
    sweepServo.write(servoAngle);
    delay(15);
  }
  checkObstacle();

  // Move back to left side
  for (servoAngle = 90; servoAngle >= 30; servoAngle--) {
    sweepServo.write(servoAngle);
    delay(15);
  }
  checkObstacle();
}

// -------------------- Distance measurement & LED logic --------------------
```

```cpp
void checkObstacle() {
  long echoDuration;
  long distanceCm;

  // Prepare ultrasonic pin for trigger pulse
  pinMode(ultraPin, OUTPUT);
  digitalWrite(ultraPin, LOW);
  delayMicroseconds(2);

  // Send ultrasonic trigger signal
  digitalWrite(ultraPin, HIGH);
  delayMicroseconds(5);
  digitalWrite(ultraPin, LOW);

  // Switch pin to input mode to read echo
  pinMode(ultraPin, INPUT);
  echoDuration = pulseIn(ultraPin, HIGH);

  // Convert echo time to distance
  distanceCm = echoTimeToCm(echoDuration);

  // Display distance on Serial Monitor
  Serial.print("Detected distance: ");
  Serial.print(distanceCm);
  Serial.println(" cm");

  // LED indication based on distance threshold
  if (distanceCm < 100) {
    digitalWrite(ledDanger, HIGH);
    digitalWrite(ledSafe, LOW);
  } else {
    digitalWrite(ledDanger, LOW);
    digitalWrite(ledSafe, HIGH);
  }
}
```

```
// ------------------- Main loop -------------------
void loop() {
  performSweep();   // execute full scan
  delay(100);       // short pause before next scan
}

// ------------------- Time-to-distance conversion -------------------
long echoTimeToCm(long echoMicroseconds) {
  return echoMicroseconds / 29 / 2;
}
```

Simulators for connecting servomotor and ultrasonic sensors to robots - 2

This program is similar to the program written earlier. However, instead of two LED; three LEDs have been used in this program. These are red, blue and orange. If there is any obstruction in front of the sensor, blue will turn on and red & orange lead will turn on for right and left obstruction.

This program can also be used for simulation only.

```
#include <Servo.h>   // Microcontroller Servo library

// -------------------- Pin setup --------------------
// Ultrasonic sensor uses a single I/O pin (trigger + echo on same line)
const int ultraIOPin  = 7;

// Indicator LEDs for three scan zones
const int ledLeftPin   = 2;   // left zone LED
const int ledCenterPin = 3;   // center zone LED
const int ledRightPin  = 4;   // right zone LED

// -------------------- Servo object --------------------
Servo headServo;          // servo that rotates the sensor

// -------------------- State variables --------------------
int scanAngle = 30;       // current servo angle (degrees)
int inchDummy = 0;        // reserved (unused)
int cmDummy   = 0;        // reserved (unused)

// Flags storing obstacle detection results (0 = clear, 1 = blocked)
int hitLeft   = 0;
int hitCenter = 0;
int hitRight  = 0;

// -------------------- Helper: distance measurement --------------------
```

```cpp
// Measures distance once and returns it in centimeters
long readDistanceCm() {
  long pulseTime;          // echo pulse length in microseconds
  long distCm;             // computed distance in centimeters

  // Set pin as OUTPUT to send trigger pulse
  pinMode(ultraIOPin, OUTPUT);
  digitalWrite(ultraIOPin, LOW);     // stabilize low state
  delayMicroseconds(2);

  // Send short HIGH trigger pulse
  digitalWrite(ultraIOPin, HIGH);
  delayMicroseconds(5);
  digitalWrite(ultraIOPin, LOW);

  // Switch pin to INPUT to read echo return
  pinMode(ultraIOPin, INPUT);
  pulseTime = pulseIn(ultraIOPin, HIGH);

  // Convert time-of-flight into distance
  distCm = usToCm(pulseTime);

  // Print for debugging/monitoring
  Serial.print("Distance reading: ");
  Serial.print(distCm);
  Serial.println(" cm");

  return distCm;
}


// -------------------- Helper: update obstacle flags --------------------
// Checks distance and stores result into the requested zone flag
void updateZoneFlag(int &zoneFlag) {
  long cm = readDistanceCm();        // take one measurement
  zoneFlag = (cm < 100) ? 1 : 0;     // threshold check (100 cm)
}
```

```
// -------------------- Helper: LED output logic --------------------
// Drive LEDs directly from the three flags
void refreshIndicators() {
  digitalWrite(ledLeftPin,   hitLeft  ? HIGH : LOW);   // left LED
  digitalWrite(ledCenterPin, hitCenter ? HIGH : LOW);   // center LED
  digitalWrite(ledRightPin,  hitRight  ? HIGH : LOW);   // right LED
}


// -------------------- Servo scan routine --------------------
// Sweeps through angles and samples at center, right, center, left
void runScanCycle() {

  // Move from left edge (30) to center (90)
  for (scanAngle = 30; scanAngle <= 90; scanAngle++) {
    headServo.write(scanAngle);     // position servo
    delay(15);                      // small motion delay
  }
  updateZoneFlag(hitCenter);        // measure center zone
  refreshIndicators();              // update LEDs

  // Move from center (90) to right edge (150)
  for (scanAngle = 90; scanAngle <= 150; scanAngle++) {
    headServo.write(scanAngle);
    delay(15);
  }
  updateZoneFlag(hitRight);         // measure right zone
  refreshIndicators();

  // Return from right edge (150) back to center (90)
  for (scanAngle = 150; scanAngle >= 90; scanAngle--) {
    headServo.write(scanAngle);
    delay(15);
  }
  updateZoneFlag(hitCenter);        // measure center again
  refreshIndicators();
```

```
  // Move from center (90) back to left edge (30)
  for (scanAngle = 90; scanAngle >= 30; scanAngle--) {
    headServo.write(scanAngle);
    delay(15);
  }
  updateZoneFlag(hitLeft);          // measure left zone
  refreshIndicators();
}

// -------------------- Microcontroller setup --------------------
void setup() {
  headServo.attach(6);              // servo signal wire on pin 6
  Serial.begin(9600);               // serial monitor speed

  // Configure LED pins as outputs
  pinMode(ledLeftPin, OUTPUT);
  pinMode(ledCenterPin, OUTPUT);
  pinMode(ledRightPin, OUTPUT);
}

// -------------------- Microcontroller main loop --------------------
void loop() {
  runScanCycle();                   // perform one full scan sequence
  delay(100);                       // short pause before repeating
}

// -------------------- Conversion function --------------------
// Convert microseconds to centimeters (speed of sound approximation)
long usToCm(long microseconds) {
  return microseconds / 29 / 2;
}
```

Connecting the robot to the servomotor and ultrasonic sensor circuit

This program is similar to the program written earlier. However, this program can be used to control ultrasonic sensors and servomotors in real circuits.

```
#include <Servo.h>  // Servo motor control library

// -------------------- Wiring / Pin Map --------------------
// HC-SR04 ultrasonic sensor (separate TRIG and ECHO)
const int pinTrig = 7;    // TRIG output pin
const int pinEcho = 8;    // ECHO input pin

// Three indicator LEDs for scan zones
const int ledL = 2;       // Left zone indicator (red LED)
const int ledC = 3;       // Center zone indicator (blue LED)
const int ledR = 4;       // Right zone indicator (orange LED)

// -------------------- Servo --------------------
Servo sensorServo;        // servo that rotates the ultrasonic sensor

// -------------------- Variables (placeholders kept) --------------------
int angleDeg = 30;        // current scanning angle
int inchHold = 0;         // unused compatibility placeholder
int cmHold   = 0;         // unused compatibility placeholder

// -------------------- Zone detection flags --------------------
// 0 = clear, 1 = obstacle detected within threshold
int zoneLeft  = 0;
int zoneMid   = 0;
int zoneRight = 0;

// -------------------- Ultrasonic measurement --------------------
// Sends a trigger pulse and returns the measured echo time (microseconds)
```

```
unsigned long getEchoPulseUs() {
  // Make sure trigger starts LOW
  digitalWrite(pinTrig, LOW);
  delayMicroseconds(2);

  // Send trigger pulse (short HIGH)
  digitalWrite(pinTrig, HIGH);
  delayMicroseconds(5);
  digitalWrite(pinTrig, LOW);

  // Measure the time ECHO stays HIGH
  return pulseIn(pinEcho, HIGH);
}

// Converts echo time (microseconds) to distance (cm)
long echoUsToCm(unsigned long echoUs) {
  return (long)(echoUs / 29 / 2);
}

// Takes one distance reading and returns it in cm
long readRangeCm() {
  unsigned long echoUs = getEchoPulseUs(); // measure echo duration
  long cm = echoUsToCm(echoUs);          // convert to centimeters

  // Print for debugging
  Serial.print("Measured: ");
  Serial.print(cm);
  Serial.println(" cm");

  return cm;
}

// -------------------- LED update --------------------
// LEDs reflect the current zone flags
void updateLeds() {
  digitalWrite(ledL, zoneLeft ? HIGH : LOW);  // left indicator
```

```cpp
  digitalWrite(ledC, zoneMid   ? HIGH : LOW);   // center indicator
  digitalWrite(ledR, zoneRight ? HIGH : LOW);   // right indicator
}

// -------------------- Zone sampling helpers --------------------
// Reads distance and stores result (1 = obstacle, 0 = clear)
void sampleZone(int &zoneFlag) {
  long cm = readRangeCm();
  zoneFlag = (cm < 100) ? 1 : 0;              // threshold = 100 cm
}

// -------------------- Servo scan routine --------------------
// Sweep: left->center (sample center), center->right (sample right),
// right->center (sample center), center->left (sample left)
void scanAndMeasure() {

  // Move from 30 to 90 (toward center)
  for (angleDeg = 30; angleDeg <= 90; angleDeg++) {
    sensorServo.write(angleDeg);
    delay(15);
  }
  sampleZone(zoneMid);     // center sample
  updateLeds();

  // Move from 90 to 150 (toward right)
  for (angleDeg = 90; angleDeg <= 150; angleDeg++) {
    sensorServo.write(angleDeg);
    delay(15);
  }
  sampleZone(zoneRight);   // right sample
  updateLeds();

  // Move back from 150 to 90 (return to center)
  for (angleDeg = 150; angleDeg >= 90; angleDeg--) {
    sensorServo.write(angleDeg);
    delay(15);
```

```
  }
  sampleZone(zoneMid);     // center sample again
  updateLeds();

  // Move from 90 down to 30 (toward left)
  for (angleDeg = 90; angleDeg >= 30; angleDeg--) {
    sensorServo.write(angleDeg);
    delay(15);
  }
  sampleZone(zoneLeft);    // left sample
  updateLeds();
}

// -------------------- Microcontroller setup --------------------
void setup() {
  sensorServo.attach(6);        // servo signal pin
  Serial.begin(9600);           // serial monitor baud rate

  // Sensor pins
  pinMode(pinTrig, OUTPUT);
  pinMode(pinEcho, INPUT);

  // LED pins
  pinMode(ledL, OUTPUT);
  pinMode(ledC, OUTPUT);
  pinMode(ledR, OUTPUT);
}

// -------------------- Main loop --------------------
void loop() {
  scanAndMeasure();             // run one full scan sequence
  delay(100);                   // short pause between cycles
}
```

Simulation of integration of ultrasonic sensors, servomotors and drive motors with robots

This program is based on the same idea as the earlier one, but the hardware and the outputs have changed. Instead of using LEDs to show the obstacle status, this version uses two DC motors for movement and a servo-mounted ultrasonic sensor for detection.

In the earlier programs, the sensor readings were used to switch indicator LEDs on and off to represent whether an obstacle was present. In this program, the sensor readings are used to control the motors directly. Depending on where an obstacle is detected around the robot, the two motors will rotate forward, backward, or in opposite directions so the robot can change its path and avoid collisions.

Like the previous version, this program is written for simulation purposes only, so it may require small adjustments before being used in the physical circuit.

```cpp
#include <Servo.h>

//
=======================================================================
==
// Hardware pins
//
=======================================================================
==

// Ultrasonic sensor (single-pin mode: shared trigger/echo)
const int ULTRA_IO_PIN = 7;

// LEDs (optional indicators)
const int LED_LEFT   = 2;   // red
```

```cpp
const int LED_CENTER = 3;   // blue
const int LED_RIGHT  = 4;   // orange

// Motor driver pins (two motors)
const int M1_IN_A = 8;
const int M1_IN_B = 9;
const int M2_IN_A = 12;
const int M2_IN_B = 13;
const int M1_EN   = 10;
const int M2_EN   = 11;

// Servo pin
const int SERVO_PIN = 6;

//
================================================================
==
// Tuning constants
//
================================================================
==

const int ANGLE_LEFT   = 30;
const int ANGLE_CENTER = 90;
const int ANGLE_RIGHT  = 150;

const int OBSTACLE_CM_THRESHOLD = 100;

// pulseIn timeout prevents "freezing" if no echo is received (in microseconds)
const unsigned long PULSE_TIMEOUT_US = 25000UL; // ~4m max in air, safe
for indoors

// Motion timing (ms)
const unsigned long DRIVE_STEP_MS = 500;   // short steps are safer than
2000ms blocks
const unsigned long TURN_STEP_MS = 450;
```

```cpp
const unsigned long BRAKE_MS      = 100;

//
======================================================================
==
// Globals
//
======================================================================
==

Servo scanner;

// Zone flags (0 = clear, 1 = blocked)
int blockedLeft   = 0;
int blockedCenter = 0;
int blockedRight  = 0;

//
======================================================================
==
// Ultrasonic: measure distance in cm (single-pin trigger/echo)
//
======================================================================
==

unsigned long readEchoPulseUs() {
  // Send trigger pulse
  pinMode(ULTRA_IO_PIN, OUTPUT);
  digitalWrite(ULTRA_IO_PIN, LOW);
  delayMicroseconds(2);

  digitalWrite(ULTRA_IO_PIN, HIGH);
  delayMicroseconds(10);          // 10us is typical for many modules
  digitalWrite(ULTRA_IO_PIN, LOW);

  // Read echo pulse
```

```cpp
  pinMode(ULTRA_IO_PIN, INPUT);
  return pulseIn(ULTRA_IO_PIN, HIGH, PULSE_TIMEOUT_US);
}

long echoUsToCm(unsigned long echoUs) {
  if (echoUs == 0) {
    // Timeout (no echo): treat as "very far"
    return 999;
  }
  return (long)(echoUs / 29 / 2);
}

long measureDistanceCm() {
  unsigned long echoUs = readEchoPulseUs();
  long cm = echoUsToCm(echoUs);

  Serial.print("Distance: ");
  Serial.print(cm);
  Serial.println(" cm");

  return cm;
}

//
==================================================================
==
// Motor helpers
//
==================================================================
==

void motorsEnable(bool on) {
  digitalWrite(M1_EN, on ? HIGH : LOW);
  digitalWrite(M2_EN, on ? HIGH : LOW);
}
```

```
void motorsStopCoast() {
  // Coast stop (all LOW)
  digitalWrite(M1_IN_A, LOW);
  digitalWrite(M1_IN_B, LOW);
  digitalWrite(M2_IN_A, LOW);
  digitalWrite(M2_IN_B, LOW);
}

void driveForward() {
  digitalWrite(M1_IN_A, HIGH); digitalWrite(M1_IN_B, LOW);
  digitalWrite(M2_IN_A, HIGH); digitalWrite(M2_IN_B, LOW);
}

void driveBackward() {
  digitalWrite(M1_IN_A, LOW);  digitalWrite(M1_IN_B, HIGH);
  digitalWrite(M2_IN_A, LOW);  digitalWrite(M2_IN_B, HIGH);
}

// Gentle turns (one motor moves, other coasts)
void turnLeftForward() {
  // left motor moves, right motor coasts
  digitalWrite(M1_IN_A, HIGH); digitalWrite(M1_IN_B, LOW);
  digitalWrite(M2_IN_A, LOW);  digitalWrite(M2_IN_B, LOW);
}

void turnRightForward() {
  // right motor moves, left motor coasts
  digitalWrite(M1_IN_A, LOW);  digitalWrite(M1_IN_B, LOW);
  digitalWrite(M2_IN_A, HIGH); digitalWrite(M2_IN_B, LOW);
}

void turnLeftBackward() {
  digitalWrite(M1_IN_A, LOW);  digitalWrite(M1_IN_B, HIGH);
  digitalWrite(M2_IN_A, LOW);  digitalWrite(M2_IN_B, LOW);
}
```

```
void turnRightBackward() {
  digitalWrite(M1_IN_A, LOW);  digitalWrite(M1_IN_B, LOW);
  digitalWrite(M2_IN_A, LOW);  digitalWrite(M2_IN_B, HIGH);
}

void brakeBriefly() {
  motorsStopCoast();
  delay(BRAKE_MS);
}

//
=================================================================
==
// LED indicators (optional)
//
=================================================================
==

void updateLeds() {
  digitalWrite(LED_LEFT,   blockedLeft   ? HIGH : LOW);
  digitalWrite(LED_CENTER, blockedCenter ? HIGH : LOW);
  digitalWrite(LED_RIGHT,  blockedRight  ? HIGH : LOW);
}

//
=================================================================
==
// Scanning: sample three zones
//
=================================================================
==

void goToAngleAndSettle(int angle) {
  scanner.write(angle);
  delay(150); // allow servo + sensor to settle before measuring
}
```

```
void scanThreeZones() {
  // Center
  goToAngleAndSettle(ANGLE_CENTER);
  blockedCenter = (measureDistanceCm() < OBSTACLE_CM_THRESHOLD) ?
1 : 0;
  updateLeds();

  // Right
  goToAngleAndSettle(ANGLE_RIGHT);
  blockedRight = (measureDistanceCm() < OBSTACLE_CM_THRESHOLD) ? 1
: 0;
  updateLeds();

  // Left
  goToAngleAndSettle(ANGLE_LEFT);
  blockedLeft = (measureDistanceCm() < OBSTACLE_CM_THRESHOLD) ? 1 :
0;
  updateLeds();

  // Return to center for driving
  goToAngleAndSettle(ANGLE_CENTER);
}

//
================================================================
==
// Decision logic (simple obstacle avoidance)
//
================================================================
==

void decideAndMove() {
  // If center is blocked, back up and pivot away from the more blocked side
  if (blockedCenter) {
    brakeBriefly();
```

```
driveBackward();
delay(TURN_STEP_MS);
brakeBriefly();

// Choose turn based on side readings
if (blockedLeft && !blockedRight) {
  turnRightForward();
} else if (blockedRight && !blockedLeft) {
  turnLeftForward();
} else {
  // both blocked or both clear -> pick a default pivot
  turnLeftForward();
}
delay(TURN_STEP_MS);
brakeBriefly();
return;
}

// If center is clear but one side is blocked, steer away slightly
if (blockedLeft && !blockedRight) {
  turnRightForward();
  delay(TURN_STEP_MS);
  brakeBriefly();
  return;
}

if (blockedRight && !blockedLeft) {
  turnLeftForward();
  delay(TURN_STEP_MS);
  brakeBriefly();
  return;
}

// Otherwise go forward in short steps
driveForward();
delay(DRIVE_STEP_MS);
```

```
  brakeBriefly();
}

//
==================================================================
==
// Setup / Loop
//
==================================================================
==

void setup() {
  Serial.begin(9600);

  // LEDs
  pinMode(LED_LEFT, OUTPUT);
  pinMode(LED_CENTER, OUTPUT);
  pinMode(LED_RIGHT, OUTPUT);

  // Motor pins
  pinMode(M1_IN_A, OUTPUT);
  pinMode(M1_IN_B, OUTPUT);
  pinMode(M2_IN_A, OUTPUT);
  pinMode(M2_IN_B, OUTPUT);
  pinMode(M1_EN, OUTPUT);
  pinMode(M2_EN, OUTPUT);

  motorsEnable(true);
  motorsStopCoast();

  // Servo
  scanner.attach(SERVO_PIN);
  scanner.write(ANGLE_CENTER);
  delay(300);
}
```

```
void loop() {
  scanThreeZones();   // update blockedLeft/Center/Right
  decideAndMove();    // act on those readings
}
```

Integration of ultrasonic sensors, servomotors and drive motors with robots

This program is similar to the previous program. The difference between this program and the previous one is that it has two DC motors connected to servomotors, ultrasonic sensors instead of LEDs. In previous programs, there were three LED on-off depending on the obstacles around the robot. Here the two DC motors will continue to rotate in different directions depending on the obstacle located around the robot; As if the robot is able to avoid obstacles. This program can be used to build real circuits for robots.

```
#include <Servo.h>

// ==========================================================
// Ultrasonic (single-pin mode): TRIG and ECHO share pin
// ==========================================================
const int ULTRA_PIN = 7;     // shared trigger/echo pin


// ==========================================================
// LED indicators
// ==========================================================
const int LED_L = 2;        // red = left obstacle
const int LED_M = 3;        // blue = middle obstacle
const int LED_R = 4;        // orange = right obstacle


// ==========================================================
// Motor driver pins
// ==========================================================
const int M1A = 8;        // Motor A direction pin 1
const int M1B = 9;        // Motor A direction pin 2
const int M2A = 12;       // Motor B direction pin 1
const int M2B = 13;       // Motor B direction pin 2
const int EN_A = 10;      // enable for motor A
const int EN_B = 11;      // enable for motor B
```

```
// ================================================
// Servo scanning
// ================================================
const int SERVO_PIN = 6;
const int ANGLE_LEFT   = 30;
const int ANGLE_MID    = 90;
const int ANGLE_RIGHT  = 150;

Servo scanner;

// ================================================
// Tuning values
// ================================================
const int OBSTACLE_CM = 100;                // threshold distance
const unsigned long ECHO_TIMEOUT_US = 25000; // prevents pulseIn lockup

// ================================================
// Flags for each direction (0=clear, 1=obstacle)
// ================================================
int blockLeft  = 0;
int blockMid   = 0;
int blockRight = 0;

// ------------------------------------------------
// Convert echo time to distance in cm
// ------------------------------------------------
long usToCm(unsigned long us) {
  return (long)(us / 29UL / 2UL);
}

// ------------------------------------------------
// Send trigger + read echo duration (single pin)
// ------------------------------------------------
unsigned long readEchoUs() {
  // send trigger pulse
  pinMode(ULTRA_PIN, OUTPUT);
```

```cpp
  digitalWrite(ULTRA_PIN, LOW);
  delayMicroseconds(2);

  digitalWrite(ULTRA_PIN, HIGH);
  delayMicroseconds(10);        // 10us is typical trigger time
  digitalWrite(ULTRA_PIN, LOW);

  // read echo pulse
  pinMode(ULTRA_PIN, INPUT);
  return pulseIn(ULTRA_PIN, HIGH, ECHO_TIMEOUT_US);
}

// ------------------------------------------------------
// Read distance (cm). If timeout occurs, return "far".
// ------------------------------------------------------
long readDistanceCm() {
  unsigned long echoUs = readEchoUs();

  if (echoUs == 0) {
    // No echo detected within timeout -> treat as far away
    return 999;
  }

  return usToCm(echoUs);
}

// ------------------------------------------------------
// Sample a zone: measure distance and store flag (0/1)
// ------------------------------------------------------
void sampleZone(int &zoneFlag, const char *label) {
  long cm = readDistanceCm();

  Serial.print(label);
  Serial.print(": ");
  Serial.print(cm);
  Serial.println(" cm");
```

```
      zoneFlag = (cm < OBSTACLE_CM) ? 1 : 0;
}


// -----------------------------------------------------
// Update LEDs based on zone flags
// -----------------------------------------------------
void updateIndicators() {
  digitalWrite(LED_L, blockLeft  ? HIGH : LOW);
  digitalWrite(LED_M, blockMid   ? HIGH : LOW);
  digitalWrite(LED_R, blockRight ? HIGH : LOW);
}


// -----------------------------------------------------
// Motor routines (same behavior as your original)
// -----------------------------------------------------
void move_forward() {
  digitalWrite(M1A, HIGH); digitalWrite(M1B, LOW);
  digitalWrite(M2A, HIGH); digitalWrite(M2B, LOW);
}


void move_backward() {
  digitalWrite(M1A, LOW);  digitalWrite(M1B, HIGH);
  digitalWrite(M2A, LOW);  digitalWrite(M2B, HIGH);
}


void turn_left_forward() {
  digitalWrite(M1A, HIGH); digitalWrite(M1B, LOW);
  digitalWrite(M2A, LOW);  digitalWrite(M2B, LOW);
}


void turn_right_forward() {
  digitalWrite(M1A, LOW);  digitalWrite(M1B, LOW);
  digitalWrite(M2A, HIGH); digitalWrite(M2B, LOW);
}
```

```
void turn_left_backward() {
  digitalWrite(M1A, LOW);  digitalWrite(M1B, HIGH);
  digitalWrite(M2A, LOW);  digitalWrite(M2B, LOW);
}

void turn_right_backward() {
  digitalWrite(M1A, LOW);  digitalWrite(M1B, LOW);
  digitalWrite(M2A, LOW);  digitalWrite(M2B, HIGH);
}

void brake() {
  digitalWrite(M1A, LOW); digitalWrite(M1B, LOW);
  digitalWrite(M2A, LOW); digitalWrite(M2B, LOW);
}

// ------------------------------------------------------
// Decision logic (keeps your original style)
// Note: long 2000ms delays make the robot sluggish.
// ------------------------------------------------------
void decideAndDrive() {
  // If obstacle in the middle -> stop and back-left
  if (blockMid == 1) {
    brake();
    delay(100);
    turn_left_backward();
    delay(2000);
  } else {
    move_forward();
    delay(2000);
  }

  // If obstacle on left -> turn right, else go forward
  if (blockLeft == 1) {
    turn_right_forward();
    delay(2000);
  } else {
```

```
  move_forward();
  delay(2000);
  }

  // If obstacle on right -> turn left, else go forward
  if (blockRight == 1) {
   turn_left_forward();
   delay(2000);
  } else {
   move_forward();
   delay(2000);
  }
}


// -------------------------------------------------------
// Servo scan routine: move + sample at key angles
// -------------------------------------------------------
void scanAndUpdate() {
  // Go to middle
  scanner.write(ANGLE_MID);
  delay(250);
  sampleZone(blockMid, "MID");
  updateIndicators();

  // Go to right
  scanner.write(ANGLE_RIGHT);
  delay(250);
  sampleZone(blockRight, "RIGHT");
  updateIndicators();

  // Back to middle
  scanner.write(ANGLE_MID);
  delay(250);
  sampleZone(blockMid, "MID");
  updateIndicators();
```

```
  // Go to left
  scanner.write(ANGLE_LEFT);
  delay(250);
  sampleZone(blockLeft, "LEFT");
  updateIndicators();

  // Return to middle for driving
  scanner.write(ANGLE_MID);
  delay(200);
}

// ===============================================================
// Setup / loop
// ===============================================================
void setup() {
  Serial.begin(9600);

  // Servo
  scanner.attach(SERVO_PIN);
  scanner.write(ANGLE_MID);

  // LEDs
  pinMode(LED_L, OUTPUT);
  pinMode(LED_M, OUTPUT);
  pinMode(LED_R, OUTPUT);

  // Motors
  pinMode(M1A, OUTPUT);
  pinMode(M1B, OUTPUT);
  pinMode(M2A, OUTPUT);
  pinMode(M2B, OUTPUT);
  pinMode(EN_A, OUTPUT);
  pinMode(EN_B, OUTPUT);

  // Enable motors
  digitalWrite(EN_A, HIGH);
```

```
  digitalWrite(EN_B, HIGH);

  brake(); // start safe
}

void loop() {
  scanAndUpdate();     // update left/mid/right flags
  decideAndDrive();    // act based on flags
  delay(100);
}
```

Connecting the LDR sensor to the robot

An LDR sensor is attached to the Microcontroller in this program. LDR is the light controlled register. When light falls on this LDR, the value of its resistor changes and the change in the value of this resistor can be determined with the help of Microcontroller's analog pin.

The program is written in below.

```
// stores the current reading from the light sensor
int sensorValue = 0;

void setup() {
  // configure analog pin for light sensor input
  pinMode(A0, INPUT);

  // start serial communication to observe readings
  Serial.begin(9600);

  // configure LED control pin as output (PWM capable)
  pinMode(9, OUTPUT);
}

void loop() {
  // get analog data from the light-dependent resistor (LDR)
  sensorValue = analogRead(A0);

  // print the raw sensor value to Serial Monitor
  Serial.print("Light level: ");
  Serial.println(sensorValue);

  // convert sensor range (0–1023) to PWM range (0–255)
  int ledBrightness = map(sensorValue, 0, 1023, 0, 255);

  // set LED brightness according to ambient light
```

```
analogWrite(9, ledBrightness);

// small delay to smooth output updates
delay(100);
}
```

An infra- ray sensor integration simulation with the robot

In this program we will simulate an infra-ray sensor. These sensors transmit rays that are invisible to the naked eye, and when these rays are reflected by an object, the receiver circuit of the sensor detects the reflected beams. The distance of the object from the sensor is measured by measuring the difference in the time of sending and receiving of infra rays.

The program is given in below.

```
// IR-based LED indicator program
// Turns an LED ON or OFF depending on infrared sensor input

int indicatorLed = 13;    // digital pin connected to the LED
int irValue = 0;          // stores infrared sensor reading

void setup() {
  // set LED pin as output
  pinMode(indicatorLed, OUTPUT);

  // initialize serial communication for monitoring
  Serial.begin(9600);
}

void loop() {
  // read analog input from IR sensor (0–1023 range)
  irValue = analogRead(A0);

  // print sensor value to Serial Monitor
  Serial.print("IR Sensor Value: ");
  Serial.println(irValue);

  // compare sensor value with predefined threshold
  // adjust this value according to sensor sensitivity and environment
  if (irValue >= 82) {
    // object or surface detected → turn LED ON
```

```
  digitalWrite(indicatorLed, HIGH);
} else {
  // no significant reflection → turn LED OFF
  digitalWrite(indicatorLed, LOW);
}

// brief delay to ensure stable readings
delay(10);
}
```

Interface Infra Ray sensor with Robot for real circuit

This program will be used in real circuits with infra ray sensors connected to the robot.

```
// Infrared sensor based LED control
// LED turns ON when IR sensor output is HIGH

const int irSensorPin = A0;   // IR sensor signal connected to analog pin A0
const int statusLed   = 13;   // onboard LED (or external LED)

void setup() {
  // Set LED pin as output
  pinMode(statusLed, OUTPUT);

  // Configure IR sensor pin as input
  pinMode(irSensorPin, INPUT);
}

void loop() {
  // Read the digital output from the IR sensor
  int irStatus = digitalRead(irSensorPin);

  // Control LED based on sensor state
  if (irStatus == HIGH) {
    // Object detected / signal active
    digitalWrite(statusLed, HIGH);
  } else {
    // No detection / signal inactive
    digitalWrite(statusLed, LOW);
  }
}
```

Multiple Infrared ray sensors interfacing with Microcontroller simulation

This program is similar to the above program. This program uses multiple infra-resonant sensors instead of one. This program can only be used for simulation work.

```
// Six infrared sensors driving six indicator LEDs

// ---------------- Pin definitions ----------------
const int ledPins[6] = {2, 3, 4, 5, 6, 7};      // LED output pins
const int sensorPins[6] = {A0, A1, A2, A3, A4, A5}; // IR sensor input pins

// Array to store sensor readings
int irReadings[6];

const int THRESHOLD = 82;   // detection threshold for IR sensors

void setup() {
  // Configure all LED pins as outputs
  for (int i = 0; i < 6; i++) {
    pinMode(ledPins[i], OUTPUT);
  }

  // Start serial communication for monitoring
  Serial.begin(9600);
}

void loop() {
  // Read all IR sensors
  for (int i = 0; i < 6; i++) {
    irReadings[i] = analogRead(sensorPins[i]);
  }

  // Display sensor values on Serial Monitor
```

```
for (int i = 0; i < 6; i++) {
  Serial.print("Sensor ");
  Serial.print(i + 1);
  Serial.print(": ");
  Serial.println(irReadings[i]);
}

// Control LEDs based on sensor threshold
for (int i = 0; i < 6; i++) {
  if (irReadings[i] >= THRESHOLD) {
    digitalWrite(ledPins[i], HIGH);   // object detected
  } else {
    digitalWrite(ledPins[i], LOW);    // no detection
  }
}

// Short delay for stable operation
delay(10);
}
```

Multiple Infrared ray sensors interfacing with Microcontroller Circuit

This program is similar to the above program. This program uses multiple infra-resonant sensors instead of one. This program can be used to build real circuits.

```
// -------- IR Sensor Pins --------
const int irSensor1 = A0;
const int irSensor2 = A1;
const int irSensor3 = A2;
const int irSensor4 = A3;
const int irSensor5 = A4;
const int irSensor6 = A5;

// -------- LED Pins --------
const int led1 = 2;
const int led2 = 3;
const int led3 = 4;
const int led4 = 5;
const int led5 = 6;
const int led6 = 7;

void setup() {
  // Configure LED pins as outputs
  pinMode(led1, OUTPUT);
  pinMode(led2, OUTPUT);
  pinMode(led3, OUTPUT);
  pinMode(led4, OUTPUT);
  pinMode(led5, OUTPUT);
  pinMode(led6, OUTPUT);

  // Configure IR sensor pins as inputs
  pinMode(irSensor1, INPUT);
  pinMode(irSensor2, INPUT);
  pinMode(irSensor3, INPUT);
```

```
  pinMode(irSensor4, INPUT);
  pinMode(irSensor5, INPUT);
  pinMode(irSensor6, INPUT);
}

void loop() {
  digitalWrite(led1, digitalRead(irSensor1));
  digitalWrite(led2, digitalRead(irSensor2));
  digitalWrite(led3, digitalRead(irSensor3));
  digitalWrite(led4, digitalRead(irSensor4));
  digitalWrite(led5, digitalRead(irSensor5));
  digitalWrite(led6, digitalRead(irSensor6));
}
```

Obstacle, Edge detection and avoidance program for Robot simulation

Using the ultrasonic sensor mounted at the front, the robot can detect obstacles ahead, as well as scan to the right and left. To improve awareness around the rest of the body, additional infrared sensors are also installed so the robot can sense nearby objects and identify dangerous edges.

In total, six infrared sensors are used on this robot. Two of these are positioned to monitor the rear-left and rear-right areas, helping the robot notice obstacles behind it on either side. The remaining four infrared sensors are placed near the edges of the chassis to detect table boundaries. Their purpose is to prevent the robot from rolling off a table or similar surface.

These edge sensors work using reflected signals. When the sensor emits its signal and receives a reflection back, the robot assumes there is a surface underneath and it is safe to continue. If the reflection does not return, it usually means the surface has ended and the robot is close to the edge.

In this program, we will use the infrared sensors to identify all unsafe directions and areas around the robot so it can avoid them during movement. This program is intended for simulation use only.

```
// Obstacle Map Generation Program for Microcontroller Table Robot (fixed &
cleaned)
#include <Servo.h>

// ---------------- Ultrasonic (single pin) ----------------
const int pingPin = 7;   // shared trigger/echo pin

// ---------------- LEDs (optional indicators) -------------
const int red_ledPin  = 2;
const int blue_ledPin = 3;
```

```
const int orng_ledPin = 4;

// ---------------- IR sensor pins -------------------------
// NOTE: If these are DIGITAL-output IR modules, digitalRead() is correct.
// If they output analog values, you must use analogRead() instead.
const int IRSensor1 = A5;
const int IRSensor2 = A4;
const int IRSensor3 = A3;
const int IRSensor4 = A2;
const int IRSensor5 = A1;
const int IRSensor6 = A0;

// ---------------- Motor driver pins ----------------------
#define INPUT1  8
#define INPUT2  9
#define INPUT3  12
#define INPUT4  13
#define ENABLE1 10
#define ENABLE2 11

// ---------------- Servo + scan settings ------------------
Servo myservo;
int pos = 30;  // servo angle

const int ANGLE_LEFT  = 30;
const int ANGLE_MID   = 90;
const int ANGLE_RIGHT = 150;

const int OBSTACLE_THRESHOLD_CM = 100;
const unsigned long PULSE_TIMEOUT_US = 25000UL; // prevents pulseIn
lockup

// ---------------- Flags ----------------
// Ultrasonic front constraints
int front_left = 0;
int front_mid  = 0;
```

```cpp
int front_right = 0;

// Edge / slope flags from IR sensors
int front_edge_right = 0;
int front_edge_left  = 0;
int back_edge_right  = 0;
int back_edge_left   = 0;

// Reverse obstacle flags
int back_obs_right = 0;
int back_obs_left  = 0;

// ----------------------------------------------------------
// Convert microseconds to centimeters
// ----------------------------------------------------------
long microsecondsToCentimeters(long microseconds) {
  return microseconds / 29 / 2;
}

// ----------------------------------------------------------
// Ultrasonic measurement helper (single-pin trigger+echo)
// ----------------------------------------------------------
long readDistanceCm() {
  // Trigger pulse
  pinMode(pingPin, OUTPUT);
  digitalWrite(pingPin, LOW);
  delayMicroseconds(2);
  digitalWrite(pingPin, HIGH);
  delayMicroseconds(10);
  digitalWrite(pingPin, LOW);

  // Echo read
  pinMode(pingPin, INPUT);
  unsigned long duration = pulseIn(pingPin, HIGH, PULSE_TIMEOUT_US);

  // If timeout -> treat as far away
```

```cpp
  if (duration == 0) return 999;

  return microsecondsToCentimeters(duration);
}

// ----------------------------------------------------------
// Obstacle checks for mid/left/right (ultrasonic)
// ----------------------------------------------------------
void obstacle_mid() {
  long cm = readDistanceCm();

  Serial.print("MID Distance: ");
  Serial.print(cm);
  Serial.println(" cm");

  front_mid = (cm < OBSTACLE_THRESHOLD_CM) ? 1 : 0;
}

void obstacle_left() {
  long cm = readDistanceCm();

  Serial.print("LEFT Distance: ");
  Serial.print(cm);
  Serial.println(" cm");

  front_left = (cm < OBSTACLE_THRESHOLD_CM) ? 1 : 0;
}

void obstacle_right() {
  long cm = readDistanceCm();

  Serial.print("RIGHT Distance: ");
  Serial.print(cm);
  Serial.println(" cm");

  front_right = (cm < OBSTACLE_THRESHOLD_CM) ? 1 : 0;
```

```
}

// ---------------------------------------------------------
// IR Check Function (updates edge + rear obstacle flags)
// ---------------------------------------------------------
void ir_check() {
  back_edge_left  = (digitalRead(IRSensor1) == HIGH) ? 1 : 0;
  back_edge_right = (digitalRead(IRSensor2) == HIGH) ? 1 : 0;

  back_obs_left   = (digitalRead(IRSensor3) == HIGH) ? 1 : 0;
  back_obs_right  = (digitalRead(IRSensor4) == HIGH) ? 1 : 0;

  front_edge_left  = (digitalRead(IRSensor5) == HIGH) ? 1 : 0;

  // BUG FIX: original code set front_edge_left in the else block (wrong)
  front_edge_right = (digitalRead(IRSensor6) == HIGH) ? 1 : 0;
}

// ---------------------------------------------------------
// Servo sweep routine (scan + update flags)
// ---------------------------------------------------------
void servo_sweep() {
  // Sweep 30 -> 90 (mid)
  for (pos = ANGLE_LEFT; pos <= ANGLE_MID; pos += 1) {
    myservo.write(pos);
    delay(15);
  }
  obstacle_mid();
  ir_check();
  control();

  // Sweep 90 -> 150 (right)
  for (pos = ANGLE_MID; pos <= ANGLE_RIGHT; pos += 1) {
    myservo.write(pos);
    delay(15);
  }
```

```
  obstacle_right();
  ir_check();
  control();

  // Sweep 150 -> 90 (mid)
  for (pos = ANGLE_RIGHT; pos >= ANGLE_MID; pos -= 1) {
    myservo.write(pos);
    delay(15);
  }
  obstacle_mid();
  ir_check();
  control();

  // Sweep 90 -> 30 (left)
  for (pos = ANGLE_MID; pos >= ANGLE_LEFT; pos -= 1) {
    myservo.write(pos);
    delay(15);
  }
  obstacle_left();
  ir_check();
  control();
}

// ---------------------------------------------------------
// Motor routines
// ---------------------------------------------------------
void move_forward() {
  digitalWrite(INPUT1, HIGH); digitalWrite(INPUT2, LOW);
  digitalWrite(INPUT3, HIGH); digitalWrite(INPUT4, LOW);
}

void move_backward() {
  digitalWrite(INPUT1, LOW);  digitalWrite(INPUT2, HIGH);
  digitalWrite(INPUT3, LOW);  digitalWrite(INPUT4, HIGH);
}
```

```
void turn_left_forward() {
  digitalWrite(INPUT1, HIGH); digitalWrite(INPUT2, LOW);
  digitalWrite(INPUT3, LOW);  digitalWrite(INPUT4, LOW);
}

void turn_right_forward() {
  digitalWrite(INPUT1, LOW);  digitalWrite(INPUT2, LOW);
  digitalWrite(INPUT3, HIGH); digitalWrite(INPUT4, LOW);
}

void turn_left_backward() {
  digitalWrite(INPUT1, LOW);  digitalWrite(INPUT2, HIGH);
  digitalWrite(INPUT3, LOW);  digitalWrite(INPUT4, LOW);
}

void turn_right_backward() {
  digitalWrite(INPUT1, LOW);  digitalWrite(INPUT2, LOW);
  digitalWrite(INPUT3, LOW);  digitalWrite(INPUT4, HIGH);
}

void station_turn_left() {
  digitalWrite(INPUT1, HIGH); digitalWrite(INPUT2, LOW);
  digitalWrite(INPUT3, LOW);  digitalWrite(INPUT4, HIGH);
}

void station_turn_right() {
  digitalWrite(INPUT1, LOW);  digitalWrite(INPUT2, HIGH);
  digitalWrite(INPUT3, HIGH); digitalWrite(INPUT4, LOW);
}

void brake() {
  digitalWrite(INPUT1, LOW); digitalWrite(INPUT2, LOW);
  digitalWrite(INPUT3, LOW); digitalWrite(INPUT4, LOW);
}

// ----------------------------------------------------------
```

```
// Control logic (kept close to your original intent)
// NOTE: You can improve this later (shorter steps, less delay)
// ------------------------------------------------------------
void control() {
  // Example LED status (optional): show front zones
  digitalWrite(blue_ledPin,  front_mid ? HIGH : LOW);
  digitalWrite(red_ledPin,   front_left ? HIGH : LOW);
  digitalWrite(orng_ledPin,  front_right ? HIGH : LOW);

  // FRONT ultrasonic logic
  if (front_mid == 1) {
    brake();
    turn_left_backward();
    delay(200);
  } else {
    move_forward();
    delay(200);
  }

  if (front_left == 1) {
    turn_right_forward();
    delay(200);
  } else {
    move_forward();
    delay(200);
  }

  if (front_right == 1) {
    turn_left_forward();
    delay(200);
  } else {
    move_forward();
    delay(200);
  }

  // EDGE checks (IR)
```

```
  if (front_edge_right == 1) {
   turn_left_backward();   // steer away from edge
   delay(200);
  }

  if (front_edge_left == 1) {
   turn_right_backward();  // steer away from edge
   delay(200);
  }

  if (back_edge_right == 1) {
   turn_left_forward();
   delay(200);
  }

  if (back_edge_left == 1) {
   turn_right_forward();
   delay(200);
  }

  // BACK obstacle checks (IR)
  if (back_obs_right == 1) {
   turn_left_forward();
   delay(200);
  }

  if (back_obs_left == 1) {
   turn_right_forward();
   delay(200);
  }
}

// ----------------------------------------------------------
void setup() {
 myservo.attach(6);
 Serial.begin(9600);
```

```cpp
  pinMode(red_ledPin, OUTPUT);
  pinMode(blue_ledPin, OUTPUT);
  pinMode(orng_ledPin, OUTPUT);

  // Motor setup
  pinMode(INPUT1, OUTPUT);
  pinMode(INPUT2, OUTPUT);
  pinMode(INPUT3, OUTPUT);
  pinMode(INPUT4, OUTPUT);
  pinMode(ENABLE1, OUTPUT);
  pinMode(ENABLE2, OUTPUT);

  digitalWrite(ENABLE1, HIGH);
  digitalWrite(ENABLE2, HIGH);

  // IR sensor setup
  pinMode(IRSensor1, INPUT);
  pinMode(IRSensor2, INPUT);
  pinMode(IRSensor3, INPUT);
  pinMode(IRSensor4, INPUT);
  pinMode(IRSensor5, INPUT);
  pinMode(IRSensor6, INPUT);

  brake();
  myservo.write(ANGLE_MID);
  delay(300);
}

void loop() {
  servo_sweep();
  delay(100);
}
```

Obstacle, Edge detection and avoidance program for Robot circuit

Thus in this program we will use the infra-ray sensor to identify all the places around the robot that the robot cannot go. This program can be used to build real circuits for robots.

```
// Obstacle Map Generation Program for Microcontroller Table Robot
#include <Servo.h>

// ---------------- Ultrasonic (single-pin mode) ----------------
// In single-pin mode, trigger and echo are the same pin
const int pingPin = 7;

// ---------------- LED pins (optional indicators) --------------
const int red_ledPin  = 2;
const int blue_ledPin = 3;
const int orng_ledPin = 4;

// ---------------- IR sensor pins ------------------------------
// Works with digital-output IR modules on analog-capable pins A0..A5
const int IRSensor1 = A5;
const int IRSensor2 = A4;
const int IRSensor3 = A3;
const int IRSensor4 = A2;
const int IRSensor5 = A1;
const int IRSensor6 = A0;

// ---------------- Motor driver pins ---------------------------
#define INPUT1  8
#define INPUT2  9
#define INPUT3  12
#define INPUT4  13
```

```cpp
#define ENABLE1 10
#define ENABLE2 11

// ---------------- Servo -------------------------------------
Servo myservo;
int pos = 30;

const int ANGLE_LEFT  = 30;
const int ANGLE_MID   = 90;
const int ANGLE_RIGHT = 150;

const int OBSTACLE_CM = 100;
const unsigned long PULSE_TIMEOUT_US = 25000UL; // prevents freezing

// ---------------- Flags -------------------------------------
// Forward constraints from ultrasonic scan
int front_left  = 0;
int front_mid   = 0;
int front_right = 0;

// Edge / slope sensors (IR)
int front_edge_right = 0, front_edge_left = 0;
int back_edge_right  = 0, back_edge_left  = 0;

// Reverse obstacles (IR)
int back_obs_right = 0, back_obs_left = 0;

// ------------------------------------------------------------
// Convert microseconds to centimeters
// ------------------------------------------------------------
long microsecondsToCentimeters(long microseconds) {
  return microseconds / 29 / 2;
}

// ------------------------------------------------------------
// Ultrasonic read (single-pin trigger + echo)
```

```
// returns distance in cm, 999 if timeout
// -----------------------------------------------------------------
long readDistanceCm() {
  // Trigger pulse
  pinMode(pingPin, OUTPUT);
  digitalWrite(pingPin, LOW);
  delayMicroseconds(2);
  digitalWrite(pingPin, HIGH);
  delayMicroseconds(10);
  digitalWrite(pingPin, LOW);

  // Echo pulse
  pinMode(pingPin, INPUT);
  unsigned long duration = pulseIn(pingPin, HIGH, PULSE_TIMEOUT_US);

  if (duration == 0) return 999; // no echo
  return microsecondsToCentimeters(duration);
}

// -----------------------------------------------------------------
// Ultrasonic obstacle checks
// -----------------------------------------------------------------
void obstacle_mid() {
  long cm = readDistanceCm();

  Serial.print("MID Distance: ");
  Serial.print(cm);
  Serial.println(" cm");

  front_mid = (cm < OBSTACLE_CM) ? 1 : 0;
}

void obstacle_left() {
  long cm = readDistanceCm();

  Serial.print("LEFT Distance: ");
```

```cpp
  Serial.print(cm);
  Serial.println(" cm");

  front_left = (cm < OBSTACLE_CM) ? 1 : 0;
}

void obstacle_right() {
  long cm = readDistanceCm();

  Serial.print("RIGHT Distance: ");
  Serial.print(cm);
  Serial.println(" cm");

  front_right = (cm < OBSTACLE_CM) ? 1 : 0;
}

// -------------------------------------------------------------
// IR check function (updates all IR flags)
// -------------------------------------------------------------
void ir_check() {
  back_edge_left  = (digitalRead(IRSensor1) == HIGH) ? 1 : 0;
  back_edge_right = (digitalRead(IRSensor2) == HIGH) ? 1 : 0;

  back_obs_left   = (digitalRead(IRSensor3) == HIGH) ? 1 : 0;
  back_obs_right  = (digitalRead(IRSensor4) == HIGH) ? 1 : 0;

  front_edge_left  = (digitalRead(IRSensor5) == HIGH) ? 1 : 0;

  // FIX: else must update front_edge_right, not front_edge_left
  front_edge_right = (digitalRead(IRSensor6) == HIGH) ? 1 : 0;
}

// -------------------------------------------------------------
// Motor functions
// -------------------------------------------------------------
void move_forward() {
```

```
  digitalWrite(INPUT1, HIGH); digitalWrite(INPUT2, LOW);
  digitalWrite(INPUT3, HIGH); digitalWrite(INPUT4, LOW);
}

void move_backward() {
  digitalWrite(INPUT1, LOW);  digitalWrite(INPUT2, HIGH);
  digitalWrite(INPUT3, LOW);  digitalWrite(INPUT4, HIGH);
}

void turn_left_forward() {
  digitalWrite(INPUT1, HIGH); digitalWrite(INPUT2, LOW);
  digitalWrite(INPUT3, LOW);  digitalWrite(INPUT4, LOW);
}

void turn_right_forward() {
  digitalWrite(INPUT1, LOW);  digitalWrite(INPUT2, LOW);
  digitalWrite(INPUT3, HIGH); digitalWrite(INPUT4, LOW);
}

void turn_left_backward() {
  digitalWrite(INPUT1, LOW);  digitalWrite(INPUT2, HIGH);
  digitalWrite(INPUT3, LOW);  digitalWrite(INPUT4, LOW);
}

void turn_right_backward() {
  digitalWrite(INPUT1, LOW);  digitalWrite(INPUT2, LOW);
  digitalWrite(INPUT3, LOW);  digitalWrite(INPUT4, HIGH);
}

void station_turn_left() {
  digitalWrite(INPUT1, HIGH); digitalWrite(INPUT2, LOW);
  digitalWrite(INPUT3, LOW);  digitalWrite(INPUT4, HIGH);
}

void station_turn_right() {
  digitalWrite(INPUT1, LOW);  digitalWrite(INPUT2, HIGH);
```

```
  digitalWrite(INPUT3, HIGH); digitalWrite(INPUT4, LOW);
}

void brake() {
  digitalWrite(INPUT1, LOW); digitalWrite(INPUT2, LOW);
  digitalWrite(INPUT3, LOW); digitalWrite(INPUT4, LOW);
}

// ------------------------------------------------------------
// Control logic (clean braces, same intent)
// ------------------------------------------------------------
void control() {
  // Optional LED indicators for front zones
  digitalWrite(blue_ledPin, front_mid ? HIGH : LOW);
  digitalWrite(red_ledPin,  front_left ? HIGH : LOW);
  digitalWrite(orng_ledPin, front_right ? HIGH : LOW);

  // FRONT (ultrasonic)
  if (front_mid == 1) {
    brake();
    turn_left_backward();
    delay(200);
  } else {
    move_forward();
    delay(200);
  }

  if (front_left == 1) {
    turn_right_forward();
    delay(200);
  } else {
    move_forward();
    delay(200);
  }

  if (front_right == 1) {
```

```
  turn_left_forward();
  delay(200);
} else {
  move_forward();
  delay(200);
}

// EDGE protection (IR)
if (front_edge_right == 1) { turn_left_backward();  delay(200); }
if (front_edge_left  == 1) { turn_right_backward(); delay(200); }

if (back_edge_right  == 1) { turn_left_forward();   delay(200); }
if (back_edge_left   == 1) { turn_right_forward();  delay(200); }

// Rear obstacle (IR)
if (back_obs_right == 1) { turn_left_forward();  delay(200); }
if (back_obs_left  == 1) { turn_right_forward(); delay(200); }
}

// -------------------------------------------------------------
// Servo sweep (scan + measure + control)
// -------------------------------------------------------------
void servo_sweep() {
  // Left -> Mid
  for (pos = ANGLE_LEFT; pos <= ANGLE_MID; pos += 1) {
    myservo.write(pos);
    delay(15);
  }
  obstacle_mid();
  ir_check();
  control();

  // Mid -> Right
  for (pos = ANGLE_MID; pos <= ANGLE_RIGHT; pos += 1) {
    myservo.write(pos);
    delay(15);
```

```
  }
  obstacle_right();
  ir_check();
  control();

  // Right -> Mid
  for (pos = ANGLE_RIGHT; pos >= ANGLE_MID; pos -= 1) {
    myservo.write(pos);
    delay(15);
  }
  obstacle_mid();
  ir_check();
  control();

  // Mid -> Left
  for (pos = ANGLE_MID; pos >= ANGLE_LEFT; pos -= 1) {
    myservo.write(pos);
    delay(15);
  }
  obstacle_left();
  ir_check();
  control();
}

// ------------------------------------------------------------
void setup() {
  myservo.attach(6);
  Serial.begin(9600);

  pinMode(red_ledPin, OUTPUT);
  pinMode(blue_ledPin, OUTPUT);
  pinMode(orng_ledPin, OUTPUT);

  // Motor setup
  pinMode(INPUT1, OUTPUT);
  pinMode(INPUT2, OUTPUT);
```

```
    pinMode(INPUT3, OUTPUT);
    pinMode(INPUT4, OUTPUT);
    pinMode(ENABLE1, OUTPUT);
    pinMode(ENABLE2, OUTPUT);

    digitalWrite(ENABLE1, HIGH);
    digitalWrite(ENABLE2, HIGH);

    // IR setup
    pinMode(IRSensor1, INPUT);
    pinMode(IRSensor2, INPUT);
    pinMode(IRSensor3, INPUT);
    pinMode(IRSensor4, INPUT);
    pinMode(IRSensor5, INPUT);
    pinMode(IRSensor6, INPUT);

    brake();
    myservo.write(ANGLE_MID);
    delay(300);
}

void loop() {
    servo_sweep();
    delay(100);
}
```

Interface buzzer device for sound generation

The program combines a buzzer to create sound and tone.

```
// -------------------- Buzzer Melody Player --------------------
// Plays a simple melody using a passive buzzer on pin 9.
// Notes are defined using characters: c d e f g a b C
// Use ' ' (space) for a rest (silence).

const int buzzerPin = 9;

// Melody string (each character is one note)
const char melody[] = "ccggaagffeeddc";

// Beat length for each note (must match melody length)
const int beatCount = 15;
int beats[beatCount] = {1, 1, 1, 1, 1, 1, 2,  1, 1, 1, 1, 1, 1, 2, 4};

// Tempo in milliseconds (bigger = slower)
const int tempo = 300;

// -------------------- Low-level tone generator --------------------
// Creates a square wave manually.
// "halfPeriodUs" is the delay in microseconds for each half wave.
// "durationMs" is how long to play the note.
void playTone(int halfPeriodUs, int durationMs) {
  long totalCycles = (long)durationMs * 1000L;   // convert ms -> microseconds

  for (long t = 0; t < totalCycles; t += (long)halfPeriodUs * 2L) {
    digitalWrite(buzzerPin, HIGH);
    delayMicroseconds(halfPeriodUs);
    digitalWrite(buzzerPin, LOW);
    delayMicroseconds(halfPeriodUs);
  }
}
```

```cpp
// -------------------- Convert note character to tone --------------------
void playNote(char noteChar, int durationMs) {
  // Note names and their corresponding half-period values (microseconds)
  // These values are commonly used for Microcontroller buzzer examples.
  char noteNames[] = {'c', 'd', 'e', 'f', 'g', 'a', 'b', 'C'};
  int noteTones[]  = {1915, 1700, 1519, 1432, 1275, 1136, 1014, 956};

  // Find the matching note
  for (int i = 0; i < 8; i++) {
    if (noteNames[i] == noteChar) {
      playTone(noteTones[i], durationMs);
      return; // stop searching after playing
    }
  }

  // If noteChar was not found, do nothing (silent)
}

void setup() {
  pinMode(buzzerPin, OUTPUT);
}

void loop() {
  // Play the melody once
  for (int i = 0; i < beatCount; i++) {
    int noteDuration = beats[i] * tempo;

    // If this character is a space, it means "rest"
    if (melody[i] == ' ') {
      delay(noteDuration);
    } else {
      playNote(melody[i], noteDuration);
    }

    // Small gap between notes
    delay(tempo / 2);
```

```
  }

  // Pause before repeating the whole song
  delay(1000);
}
```

Programming a table guard robot

In this program, the robot stays in one place and rotates left and right at a certain interval. At the same time, the ultrasonic sensor will continue to scan on both sides. If the sensor detects an obstruction, it will alert the intruder by making noise using the buzzer. Note that, in order for the robot to be used effectively in guard mode, it must be placed in an empty space, so that the sensor does not create an alarm as an object is thought to be intruder.

```
#include <Servo.h>

// ---------------- Ultrasonic (single-pin mode) ----------------
const int pingPin = 7;     // shared trigger/echo pin

// ---------------- LEDs ----------------
const int red_ledPin  = 2;
const int blue_ledPin = 3;
const int orng_ledPin = 4;

// ---------------- Buzzer ----------------
// Changed from 4 to 5 to avoid conflict with orange LED
const int buzzerPin = 5;

// ---------------- IR pins (not used in this code yet) --------
const int IRSensor1 = A5;
const int IRSensor2 = A4;
const int IRSensor3 = A3;
const int IRSensor4 = A2;
const int IRSensor5 = A1;
const int IRSensor6 = A0;

// ---------------- Motor driver pins -------------------------
#define INPUT1  8
#define INPUT2  9
#define INPUT3  12
#define INPUT4  13
```

```
#define ENABLE1 10
#define ENABLE2 11

// ---------------- Servo --------------------------------------
Servo myservo;
int pos = 30;

// Scan angles
const int ANGLE_LEFT  = 30;
const int ANGLE_MID   = 90;
const int ANGLE_RIGHT = 150;

const int THRESHOLD_CM = 100;
const unsigned long PULSE_TIMEOUT_US = 25000UL; // prevents freeze

// ------------------------------------------------------------
// Convert echo microseconds to centimeters
// ------------------------------------------------------------
long microsecondsToCentimeters(unsigned long microseconds) {
  return (long)(microseconds / 29UL / 2UL);
}


// ------------------------------------------------------------
// Single-pin ultrasonic read (returns distance in cm)
// ------------------------------------------------------------
long readDistanceCm() {
  // Trigger pulse
  pinMode(pingPin, OUTPUT);
  digitalWrite(pingPin, LOW);
  delayMicroseconds(2);
  digitalWrite(pingPin, HIGH);
  delayMicroseconds(10);
  digitalWrite(pingPin, LOW);

  // Echo pulse
  pinMode(pingPin, INPUT);
```

```
unsigned long duration = pulseIn(pingPin, HIGH, PULSE_TIMEOUT_US);

  if (duration == 0) return 999; // no echo -> far away
  return microsecondsToCentimeters(duration);
}

// --------------------------------------------------------------
// Check obstacle and beep if close
// --------------------------------------------------------------
void obstacleCheck() {
  long cm = readDistanceCm();

  Serial.print("Distance: ");
  Serial.print(cm);
  Serial.println(" cm");

  if (cm < THRESHOLD_CM) {
    digitalWrite(buzzerPin, HIGH);   // buzzer ON
  } else {
    digitalWrite(buzzerPin, LOW);    // buzzer OFF
  }
}

// --------------------------------------------------------------
// Servo sweep: scan and check distance at key points
// --------------------------------------------------------------
void servo_sweep() {
  // Left -> Mid
  for (pos = ANGLE_LEFT; pos <= ANGLE_MID; pos += 1) {
    myservo.write(pos);
    delay(15);
  }
  obstacleCheck();

  // Mid -> Right
  for (pos = ANGLE_MID; pos <= ANGLE_RIGHT; pos += 1) {
```

```cpp
    myservo.write(pos);
    delay(15);
  }
  obstacleCheck();

  // Right -> Mid
  for (pos = ANGLE_RIGHT; pos >= ANGLE_MID; pos -= 1) {
    myservo.write(pos);
    delay(15);
  }
  obstacleCheck();

  // Mid -> Left
  for (pos = ANGLE_MID; pos >= ANGLE_LEFT; pos -= 1) {
    myservo.write(pos);
    delay(15);
  }
  obstacleCheck();
}

// -------------------------------------------------------------
// Motor functions
// -------------------------------------------------------------
void station_turn_left() {
  digitalWrite(INPUT1, HIGH); digitalWrite(INPUT2, LOW);
  digitalWrite(INPUT3, LOW);  digitalWrite(INPUT4, HIGH);
}

void station_turn_right() {
  digitalWrite(INPUT1, LOW);  digitalWrite(INPUT2, HIGH);
  digitalWrite(INPUT3, HIGH); digitalWrite(INPUT4, LOW);
}

void brake() {
  digitalWrite(INPUT1, LOW); digitalWrite(INPUT2, LOW);
  digitalWrite(INPUT3, LOW); digitalWrite(INPUT4, LOW);
```

```
}

// -----------------------------------------------------------
void setup() {
  myservo.attach(6);
  Serial.begin(9600);

  // LEDs
  pinMode(red_ledPin, OUTPUT);
  pinMode(blue_ledPin, OUTPUT);
  pinMode(orng_ledPin, OUTPUT);

  // Buzzer
  pinMode(buzzerPin, OUTPUT);
  digitalWrite(buzzerPin, LOW);

  // Motors
  pinMode(INPUT1, OUTPUT);
  pinMode(INPUT2, OUTPUT);
  pinMode(INPUT3, OUTPUT);
  pinMode(INPUT4, OUTPUT);
  pinMode(ENABLE1, OUTPUT);
  pinMode(ENABLE2, OUTPUT);

  digitalWrite(ENABLE1, HIGH);
  digitalWrite(ENABLE2, HIGH);

  // IR sensors (configured, not used here)
  pinMode(IRSensor1, INPUT);
  pinMode(IRSensor2, INPUT);
  pinMode(IRSensor3, INPUT);
  pinMode(IRSensor4, INPUT);
  pinMode(IRSensor5, INPUT);
  pinMode(IRSensor6, INPUT);

  brake();
```

```
  myservo.write(ANGLE_MID);
  delay(300);
}

void loop() {
  // Scan and beep if obstacle is close
  servo_sweep();
  delay(100);

  // Demo: spin left then right
  station_turn_left();
  delay(1500);

  station_turn_right();
  delay(1500);

  brake();
  delay(200);
}
```

It is possible to write a new program by modifying the above program where it will scan the surrounding position for some time after launching the robot; but don't make any alarms. This will create a map of the objects around it. When the allotted time is up, it will turn on the alarm if it detects anything new except the surrounding objects.